Engineering Cyber-Physical Systems and Critical Infrastructures

Volume 12

Series Editor

Fatos Xhafa, Departament de Ciències de la Computació, Technical University of Catalonia, Barcelona, Spain

The aim of this book series is to present state of the art studies, research and best engineering practices, real-world applications and real-world case studies for the risks, security, and reliability of critical infrastructure systems and Cyber-Physical Systems. Volumes of this book series will cover modelling, analysis, frameworks, digital twin simulations of risks, failures and vulnerabilities of cyber critical infrastructures as well as will provide ICT approaches to ensure protection and avoid disruption of vital fields such as economy, utility supplies networks, telecommunications, transports, etc. in the everyday life of citizens. The intertwine of cyber and real nature of critical infrastructures will be analyzed and challenges of risks, security, and reliability of critical infrastructure systems will be revealed. Computational intelligence provided by sensing and processing through the whole spectrum of Cloud-to-thing continuum technologies will be the basis for real-time detection of risks, threats, anomalies, etc. in cyber critical infrastructures and will prompt for human and automated protection actions. Finally, studies and recommendations to policy makers, managers, local and governmental administrations and global international organizations will be sought.

Massimo Ficco · Gianni D'Angelo
Editors

Artificial Intelligence Techniques for Analysing Sensitive Data in Medical Cyber-Physical Systems

System Protection and Data Analysis

Springer

Editors
Massimo Ficco
Department of Computer Science
University of Salerno
Fisciano, Italy

Gianni D'Angelo
Department of Computer Science
University of Salerno
Fisciano, Italy

ISSN 2731-5002 ISSN 2731-5010 (electronic)
Engineering Cyber-Physical Systems and Critical Infrastructures
ISBN 978-3-031-70774-2 ISBN 978-3-031-70775-9 (eBook)
https://doi.org/10.1007/978-3-031-70775-9

This Springer imprint is published by the registered company Springer Nature Switzerland AG
The registered company address is: Gewerbestrasse 11, 6330 Cham, Switzerland

Preface

The advances in Internet of Things and Artificial Intelligence (AI) have witnessed great progress in healthcare technologies in several application domains. In particular, the interconnection between the physical space, characterized by physical devices able to collect users' health information, with the cyberspace, also known as the virtual space, has fostered the development of intelligent Medical Cyber-Physical Systems (MCPSs) with the capability to deliver real-time healthcare services. On the other hand, the potential innovation that these technologies bring to improve patient care, by remotely analysing health parameters using medical devices, advanced smart sensors, and AI, is hampered by security and privacy challenges related to the managed sensitive data.

This book presents the major advances in techniques to preserve privacy and security requirements connected with the use of AI and Machine Learning (ML) to analyse and manage sensitive data in MCPSs. Starting from the state-of-the-art on AI and ML for medical applications and digital health, an accurate analysis of privacy and security risks associated with the use of the MCPSs is presented. Then, Digital Twins are introduced as a significant technique to enhance decision-making through learning and reasoning of collected on-field real-time data. Moreover, decentralized healthcare data management approaches based on Federated Learning, Tiny Machine Learning, and blockchain technologies have been introduced to shift control and responsibility of healthcare data management from individual centralized entities to a more distributed structure, preserving privacy and security. Finally, the application of AI-based security monitoring approaches in healthcare is discussed.

In this book, both theoretical and practical approaches are used to allow readers to understand complex topics and concepts easily also through real-life scenarios. In particular, the focus and correspondingly the topics covered have been focused on techniques, technologies, and tools related to security and privacy concepts in MCPSs and the use of advanced AI and ML techniques for medical data analysis.

Fisciano (SA), Italy
June 2024

Massimo Ficco
Gianni D'Angelo

Contents

Artificial Intelligence and Machine Learning for Medical Applications and Digital Health 1
Valeria Visco, Carmine Vecchione, Gianni D'Angelo, and Michele Ciccarelli

Healthcare & Cyberspace: Medical Cyber-Physical Systems 17
Rocco Zaccagnino, Alfonso Guarino, Delfina Malandrino, and Gianluca Zaccagnino

Privacy Risk Analysis in Medical Cyber-Physical Systems 29
Michele Mastroianni

Healthcare and Cyberspace: from Cyber-Physical Systems to Medical Digital Twins 47
Rocco Zaccagnino, Delfina Malandrino, Gianluca Zaccagnino, Gerardo Benevento, and Margherita Maria Napolitano

Federated Learning in Agents Based Cyber-Physical Systems 73
Domenico Di Sivo, Palma Errico, and Salvatore Venticinque

Applications of Solid Platform and Federated Learning for Decentralized Health Data Management 95
Franco Cirillo, Marco De Santis, and Christian Esposito

TinyML and Federated Learning for Resource-Constrained Medical Devices 113
Pietro Fusco, Gennaro Pio Rimoli, and Massimo Ficco

The Role of Blockchain in AI-Driven Medical Cyber-Physical Systems 127
Biagio Boi and Christian Esposito

Artificial Intelligence Applications in Healthcare Security 143
Salvatore D'Antonio and Federica Uccello

Artificial Intelligence and Machine Learning for Medical Applications and Digital Health

Valeria Visco, Carmine Vecchione, Gianni D'Angelo, and Michele Ciccarelli

Abstract Artificial Intelligence (AI) and Machine Learning (ML) are now used in many research and industrial fields. Among these applications, they are widely employed to support medical professionals, patients, and caregivers in disease management, and they appear well-positioned to revolutionize the healthcare industry. Among the many applications of AI are medical imaging, predictive analytics, personalized treatment plans, and administrative automation. AI/ML can manage and integrate different data resources (clinical measurements and observations, biological data, experimental results, environmental information, and wearable device data) into models for managing human diseases. Incorporating AI/ML into medicine can enhance every stage of patient care, from risk stratification to diagnosis and therapy. Consequently, this chapter delves into how AI can positively affect every step of the diagnostic-therapeutic pathway, focusing on patient risk stratification, diagnosis, and disease treatment. Specifically, for each of these domains, we report the general approach of AI/ML in a real-world context. As shown, AI/ML-based approaches can improve healthcare workers' knowledge, enable them to spend more time on direct patient care, and reduce fatigue.

1 Artificial Intelligence and Machine Learning in Medical Science: Foundations and Applications

In recent years, technologies capable of performing data analysis to extract information useful for decision support have emerged. This has been made possible thanks to contributions from various disciplines, such as artificial intelligence studies, neu-

V. Visco · C. Vecchione · M. Ciccarelli (✉)
Department of Medicine, Surgery and Dentistry, University of Salerno, Fisciano, SA, Italy
e-mail: mciccarelli@unisa.it

C. Vecchione
Vascular Physiopathology Unit, IRCCS Neuromed, Pozzilli, NA, Italy

G. D'Angelo
Department of Computer Science, University of Salerno, Fisciano, SA, Italy

M. Ficco and G. D'Angelo (eds.), *Artificial Intelligence Techniques for Analysing Sensitive Data in Medical Cyber-Physical Systems*, Engineering Cyber-Physical Systems and Critical Infrastructures 12, https://doi.org/10.1007/978-3-031-70775-9_1

ral networks, statistics, and machine learning. These technologies, when applied to databases, are known by the terms Knowledge Discovery, Knowledge Extraction, Data Archaeology, and Data Mining. These terms describe, at least in part, the concept of "Knowledge Discovery in Databases" (KDD) [1], understood as the process of extracting information, or rather knowledge, from large amounts of data. Here is a commonly used definition: "*Data mining refers to the process of selecting, exploring, and modeling large amounts of data in order to discover regularities or relationships not known a priori, and with the aim of obtaining a clear and useful result for the owner of the database.*"

KDD in medicine is aimed at searching uncover insights present into medical data in order to improve health outcomes and patient experiences. Specifically, KDD-based algorithms support medical professionals in clinical settings and ongoing research. As a result, there is no doubt that AI and ML are becoming an increasingly central part of the digital health systems that support and shape modern medicine.

1.1 The Foundations of KDD Workflow

The process for KDD consists of a number of interactive stages that manipulate and transform data to extract useful information. In 1996, Fayyad et al. [2] identified five steps in the KDD process, as outlined below (Fig. 1):

- **Selection**: Raw data is segmented and selected according to certain criteria in order to arrive at a subset of data, representing the "target data" or objective data.
- **Preprocessing**: The data cleaning phase involves eliminating possible errors and deciding on mechanisms for handling missing data.
- **Transformation**: When data comes from different sources, it is necessary to reconfigure it to ensure consistency.

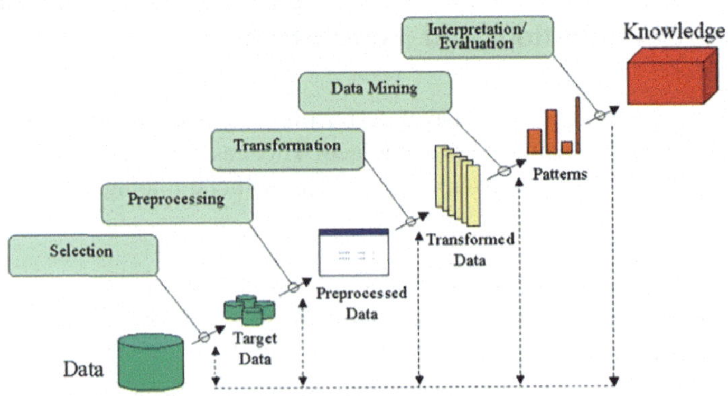

Fig. 1 Stages of the KDD process according to Fayyad et al. [2]

- **Data Mining**: Learning starts from a well-defined set of classification examples for known cases, from which one expects to deduce a way to classify unknown examples.
- **Interpretation**: The abstraction system for classification creates patterns, or models, that can provide valuable decision support. However, interpreting the results through graphs displaying the classification output is not enough; these models need to be evaluated to understand to what extent they can be useful.

1.2 Soft Computing, Machine Learning, and Artificial Intelligence

Soft Computing is the combination of methodologies aimed at modeling and enabling solutions to complex real-world problems that are either not amenable to mathematical modeling or are too complex for it. Its aspiration is to use approximation tolerance (the model's characteristics are similar to real ones but not identical), uncertainty (uncertainty exists regarding whether the model reflects reality), imprecision (the quantitative characteristics of the model are not identical to real ones but close to them), and partial truth to achieve close resemblance to human decisions.

The underlying theory of Soft Computing is to utilize tolerance capability to arrive at robust and low-cost solutions. The human mind serves as the model for Soft Computing. Some of the most well-known Soft Computing techniques include Artificial Neural Networks (ANN), Fuzzy Logic (FL), Adaptive Neuro-Fuzzy Inference Systems (ANFIS), and evolutionary computation.

However, as we will highlight in this work, the key to success for a Soft Computing-based testing system lies in choosing the right feature extraction method that accurately and uniquely represents the examined process in the most efficient manner possible.

An important aspect of Soft Computing is to determine whether computers can learn directly and autonomously from data. In this regard, Machine Learning (ML) arises from the theory that computers can learn to perform specific tasks without being explicitly programmed to do so, thanks to the recognition of patterns in data. ML utilizes algorithms that enable automatic Data Mining, learning through direct interaction with data. To this end, computers are provided with the intelligence needed to identify even unknown information without being explicitly instructed where to look for it.

The most important aspect of ML is repeatability, as the more models are exposed to data, the more they are able to adapt autonomously. Computers learn from previous computations to produce reliable and replicable results and make decisions. Thanks to new processing technologies, today's ML is not the ML of the past. This science is not new but is gaining new momentum. And while many ML algorithms have been around for a long time, the ability to apply complex mathematical calculations to big data is a more recent development. Some widely known examples of ML usage that may be familiar are:

- Autonomous driving;
- Online offer recommendations such as those from Amazon or Netflix;
- Knowing opinions about a company from customers on Twitter. Machine learning combined with the creation of linguistic rules;
- Fraud detection;

The set of algorithms and techniques (Soft-Computing) for extracting knowledge from data (Data-Mining), and the automation of this process (Machine-Learning) represent the pillars of Artificial Intelligence (AI). Therefore, in a simplistic way, we could define AI as the ability of a technological system to solve problems or perform tasks and activities typical of the human mind and ability. Looking at the computer science sector, we could identify AI as the discipline that deals with creating machines (hardware and software) capable of "acting" autonomously (solving problems, performing actions, etc.).

The typical AI/ML workflow starts with data acquisition, continues with feature engineering, and then with algorithm selection and model development, with its evaluation and application [3, 4]. Precisely, AI/ML can manage and incorporate different data resources (clinical measurements and observations, biological -omics, experimental results, environmental information, wearable devices) into models for describing and predicting human diseases [5]. The most common example is the electronic health record (EHR), which accumulates information about the patient's health from several sources and includes:

- test results (text and image data);
- clinical observations;
- current health problems;
- medications;
- previous surgical procedures.

Consequently, after data acquisition, a pre-processing step involves cleaning, normalization, and scaling to prepare data for analysis. The following ML algorithm selection is challenging: every model is unique and requires a balance between complexity and efficiency. According to the "No Free Lunch Theorem," no model perfectly works for all situations, so using as many as possible is advisable. Generally, the relationships between the components and features (different data and information recorded) in a large dataset are discovered using standard ML strategies, typically grouped into supervised and unsupervised learning. Supervised learning models use preliminary data with accurate labels for training; as a result, the model should understand the underlying class-conditional feature spaces [6]. On the other hand, Using unlabelled data, an unsupervised learning model must learn from the relationships among the elements of a dataset and classify them without basing them on categories or labels [6]. Finally, depending on the problem you have set out to solve, specific evaluation metrics assess model effectiveness and what hyperparameters might need to be tuned.

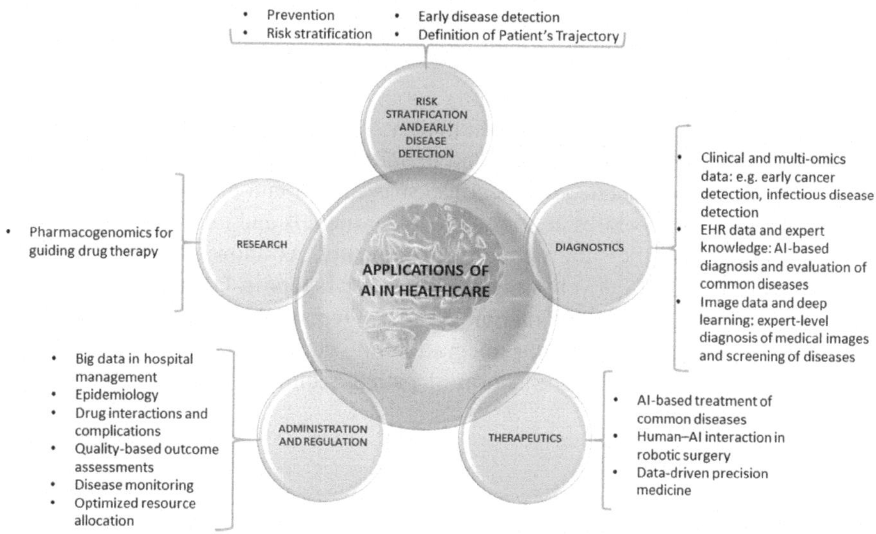

Fig. 2 Potential roles of AI in healthcare. AI is poised to play significant roles in healthcare across various application domains, including research, risk stratification/early disease detection, diagnostics, therapeutics, administration, and regulation

1.3 The Applications

There are several ways AI can positively affect the practice of medicine, whether by supporting clinicians' decision-making or speeding up the research step. Incorporating AI into medicine can improve every stage of patient care, from risk stratification to diagnosis and therapy (Fig. 2); consequently, clinical practice will become more efficient, convenient, personalized, and effective [3].

2 Patient Risk Stratification

Risk stratification allows healthcare providers to identify the right level of care and services for different subgroups of patients (of similar complexity and care needs); specifically, it assigns a risk status to patients and uses this information to direct care and increase health outcomes. Consequently, doctors must use risk stratification to improve different strategies for patients who need a programme of more specific controls.

2.1 Using AI for Patient Risk Stratification

There are numerous areas of opportunity for AI in prevention, particularly in the battle against chronic diseases [3]. Specifically, AI is essential in precision disease stratification, early disease detection, and defining a patient's trajectory. Numerous clinical risk scores are intended for specific diseases, populations, and risk groups [7]; however, most of these scores are based on linear models and might lack specificity and sensitivity in specific subgroups [8]. Specifically, AI allows the elaboration of new risk stratification to develop novel prognostic layers in favor of personalized medicine and identify different outcome profiles. Overall, the clinical advantages of implementing AI systems and their integration into current prognostic stratification might grant a satisfactory identification of patient's clinical outcomes, improving their clinical management.

2.2 Real-World Applications

AI has been applied in several settings in clinical nephrology. For instance, it has been proven helpful for predicting the decline of glomerular filtration rate in patients with polycystic kidney disease [9], or for establishing the risk for progressive IgA nephropathy [10]; as well as it has been used in gastroenterology to predict outcomes in gastrointestinal bleeding [11] and survival of esophageal cancer [12]. Among the most recent applications of AI is the possibility to help physicians and patients in risk stratification and prediction of the development of cardiovascular diseases [5]. ML can be applied to atrial fibrillation screening to support atrial fibrillation-risk prediction or improve the automated diagnosis using different rhythm monitoring modalities. Numerous clinical risk score models have been created to predict the likelihood of atrial fibrillation, mainly developed using data from specific population cohorts; however, it has been found that the performance of a model is not as precise as when performed on the original cohort [13, 14]. In recent years, technologies for heart rhythm monitoring have been improving: handheld or smartphone or smartwatch single-lead ECG recorders, ECG patch monitors, photoplethysmography smartwatches or smartphones, blood pressure monitors, and external Holter monitors. Originally, ML was applied to sinus rhythm segments of ambulatory ECGs to predict paroxysmal atrial fibrillation based mainly on premature atrial contractions; while, more recently, another approach based on heart rate variability features, including times and frequency domains, has been used with an accuracy above 0.9 [15–17]. Studies have developed ML model models to detect atrial fibrillation from a 12-lead ECG [18, 19]. Neural networks are typically used in the decision process to define if the ECG demonstrates atrial fibrillation. Neural networks warrant high accuracy values for models differentiating between atrial fibrillation and sinus rhythm; however, their performance decreased when the models were applied to determine other rhythms too. Detection of irregularities in pulse pressure signals measured peripherally supports the diagnosis of atrial fibrillation [20, 21]; however, when compared, neural networks were superior, with accuracy, specificity, and sensitivity above 0.90. Similarly, current guidelines emphasize the importance of early detection and risk stratification in coronary artery disease (CAD) to implement goal-directed medical therapies [22]. Specifically, Dogan et al. built a model of eight random-forest classifiers to predict the risk of symptomatic CAD using genetic and epigenetic variables along with clinical risk factors: the model predicted symptomatic CAD with an accuracy, sensitivity, and specificity of 0.78, 0.75, and 0.80, respectively, in the internal validation cohort [23]. Moreover, the PRAISE score, an ML-based prediction of adverse events following an acute coronary syndrome, showed accurate discriminative capabilities for predicting all-cause death, myocardial infarction, and major bleeding over a period of 1 year [24]. Finally, various ML algorithms based on stress imaging, particularly single-photon emission computed tomography, have been devised to facilitate the prediction of CAD [25, 26]. Equally, several methods emerged from the ML community that can construct predictive models in heart failure. Precisely, Aljaaf et al. [27] suggested a decision-tree model trained on the Cleveland Clinic Heart Disease Data Set, with 297 patients, each characterized by 13 clinical parameters, stratifying individuals into one of five risk classes for HF development. Model performance was assessed using 10-fold cross-validation on the training set, with a mean AUC of 0.91 and a classification accuracy of 86.5%. Moreover, rehospitalizations negatively impact healthcare systems' costs and patient loads. Consequently, the prediction of rehospitalizations has been estimated by numerous groups [28–30]; in particular, Lorenzoni et al. [31] compared the performance of 8 ML techniques in the forecast of hospitalization among HF patients, using data from the Gestione Integrata dello Scompenso Cardiaco (GISC) study: HF patients were enrolled in a longterm assistance program that includes the adoption of an online platform for data sharing between general practitioners and cardiologists. Logistic regression, generalized linear model net (GLMN), classification and regression tree, random forest, adaboost, logitboost, SVM, and ANN were used to evaluate the suitability of such methods in predicting hospitalization of 380 patients, using demographic and clinical parameters of each patient. The GLMN showed better performance in predicting hospitalization, with an average accuracy, positive predictive value, and negative predictive value of 81.2%, 87.5%, and 75%, respectively. However, just like any new drug or device, the implementation of digital medical technologies ad AI algorithms require rigorous validation with randomized, controlled clinical trials; on the contrary, at this moment research is limited by sample size necessary for inference in patient risk stratification in most medical areas.

3 The Challenge of Diagnosis

Diagnosis has been defined as both a process and a classification scheme, or a "pre-existing set of classes agreed upon by the medical profession to designate a specific condition" [32]; precisely, the goal of medical diagnosis is to provide an effective treatment. Accordingly, when a diagnosis is accurate and early, the patient has the best opportunity for a positive health outcome because clinical decision-making will be tailored to a precise understanding of the patient's disease [33].

3.1 Using AI for Disease Diagnosis

With the recent AI revolution, medical diagnostics could be improved by improving the diagnostic process's prediction accuracy, speed, and efficiency. Precisely, AI can analyze several patient data, including medical history, vital signs (e.g., body temperature, pulse rate, respiration rate, and blood pressure), medical 2D/3D imaging, bio-signals (e.g., ECG, EEG, EMG, and EHR), demographic information, and laboratory test results [34]. Consequently, by integrating multiple data sources, doctors can understand a patient's health and the underlying causes of their symptoms, reducing the chance of misdiagnosis and improving the accuracy of diagnosis. By using AI algorithms to analyze vast amounts of medical data and identify patterns and relationships, general AI can transform medicine, leading to better patient outcomes and a more efficient and effective healthcare system [34].

3.2 Real-World Applications

The specialty of gastroenterology benefits from a wide range of AI applications in diagnostic settings. Specifically, gastroenterologists used convolutional neural networks, among other deep learning models, to process images from endoscopy and ultrasound [35] and detect abnormal structures such as colonic polyps [36]. Moreover, artificial neural networks have also been used to diagnose gastroesophageal reflux disease [37] and atrophic gastritis [38]. Neurology also uses AI in the diagnostic process. Accordingly, Liu et al. [39] proposed the use of spectrogram features extracted from speech data to recognize Alzheimer's disease. The system uses the voice data collected via the Internet of Things and transmitted to the cloud server where the original data is stored. The received data is used for training the model to identify Alzheimer's disease symptoms. Similarly, Lodha et al. [40] used an ML algorithm to process the data of neuroimaging technologies to detect Alzheimer's in its primitive stage; It uses various algorithms like support vector machine (SVM), gradient boosting, K-nearest neighbor, Random forest, a neural network that shows the accuracy rate 97.56, 97.25, 95.00, 97.86, 98.36, respectively. Equally, researchers have found that AI has been able to diagnose hypertension; interestingly, Arsalan et al. [41] conducted the experiments using three publicly available datasets as digitized retinal imagery for vessel extraction (DRIVE), structured analysis of retina (STARE) for hypertension detection. They achieved accuracy for all datasets with sensitivity, specificity, area under the curve, and accuracy of 80.22%, 98.1%, 98.2%, and 96.55%, respectively. ML models are increasingly being applied to dermatology as a diagnostic decision support tool using image analysis, especially for skin cancer detection, because early skin cancer diagnosis is inevitably the cornerstone for improving mortality and morbidity outcomes. However, the standard diagnostic accuracy, even with dermoscopy, remains comparatively low [42]. Consequently, Han et al. [43] tested using a DL algorithm to classify the clinical images of 12 skin diseases, including melanoma. When tested with the validation image set, the average sensitivity and specificity for all the conditions were 85.1% and 81.3%, respectively, with an area under the receiver operating characteristic (AUROC) of 0.89. The tested algorithm performance was similar to that of 16 dermatologists. Nevertheless, the external validity of this program is still limited, as Navarrete-Dechent et al. [44], who externally tested the program in a different population, found much lower sensitivity, with the correct histopathological diagnosis identified in only 29 of the 100 lesions. In conclusion, research has focused on tasks where AI can well prove its performance in relation to a doctor. Generally, these tasks have clearly defined inputs and a binary output that is easily confirmed [45], like in classifying suspicious skin lesions, where the input is a digital photograph and the output is a simple binary classification: benign or malignant. However, most studies comparing the efficiency of AI vs. clinicians are found to have unreliable designs and are known to lack primary replication, i.e., the validation of the algorithms developed in samples coming from sources other than the one used to train algorithms in detecting diseases from medical imaging. These findings support the need for extensive validation of AI-based technologies through rigorous clinical trials. Moreover, opposing AI and clinicians is, although well represented in the scientific literature, probably not the best way to tackle the issue of performance in medical expertise: several studies are now approaching the interaction between clinicians and algorithms [46] as the combination of human and artificial intelligence outperforms either alone. Consequently, in the short term, physicians could use algorithms to aid with double-checking a diagnosis.

4 The Proper Treatment for the Right Patient

Personalized medicine is a novel and exciting topic in medicine and healthcare. It is a concept that can transform medical interventions by providing practical, tailored therapeutic strategies based on an individual's genomic, epigenomic, and proteomic profile while also remaining mindful of a patient's situation [47]. The 21st-century vision of personalized medicine is to provide "the right drug, with the right dose at the right time to the right patient" [48].

4.1 Using AI for Disease Treatment

AI algorithms can easily enhance precision medicine. Because AI models can learn and retain preferences, AI has the potential to provide customized recommendations. Rather than repeating information with a new person each time, a healthcare system could offer patients around-the-clock access to an AI-powered virtual assistant to answer questions based on the patient's medical history, preferences, and personal needs. The different AI applications allow the detection of a patient's health trajectory leading to personalized medicine and tailored treatment. Clinical trials sometimes take over a decade and are very expensive regarding pharmaceutical development. Consequently, with the help of AI, researchers can select a few clinically effective candidates out of thousands of molecules with the help of computer-aided drug design software in less time [49] whereas traditional methods may require years together. AI involves the networks which recognize some strategies in drug molecule design, and algorithms are designed that aid in understanding the differences between normal and disease profiles within a complex data set. As a result, implementing AI has reduced the time by many folds, and the development can be done within 2–4 years [50]. Moreover, AI gives surgeons a new set of eyes and a new perspective, which can introduce new methodologies to existing surgical procedures and subsequently standardize practices. Precisely, learning from thousands of different surgeries, AI-based systems can give new indications for the best surgical techniques.

4.2 Real-World Applications

In oncology, it is essential to identify the proper treatment for a patient to improve his prognosis. IBM Watson launched its unique program for oncologists to provide clinicians with evidence-based treatment options. Watsonfor Oncology can analyze the meaning and context of structured and unstructured data in clinical notes and reports that may be critical to choosing a specific treatment pathway. Then by combining parameters from the patient's file with external research, clinical expertise, and data, the program identifies potential treatment plans for a patient [51]. Moreover, among the most recent applications of AI is the possibility of helping physicians in cardiovascular disease treatment. To date, optimizing pharmacological therapy to achieve the optimal control of blood pressure levels represents a tough challenge for doctors; indeed, blood pressure control rates still need to improve worldwide [52]. In this field, AI could support the choice of the most appropriate management of arterial hypertension and can identify the best combination therapy compared to standard analysis [53]. Accordingly, Ye et al. [54] showed the potential of using predictive models to select optimal arterial hypertension treatment; specifically, along with clinical guidelines and guideline-based clinical decision support systems, the LSTM models show the best prediction for achieving the optimal BP control BP with several combinations of treatments. Similarly, many possibilities of 'big data' and ML applications exist in cardiac rehabilitation; in particular, several studies have focused on using predictive models to support doctors in choosing a specific rehabilitative treatment plan [3]. Specifically, Lofaro et al. [55] enrolled 129 patients for cardiac rehabilitation to derive a prediction model from helping cardiologists prescribe the appropriate rehabilitation program: among the different algorithms tested, the Lasso regression showed an average classification accuracy on the test set of 0.935. Finally, AI may allow identifying who will benefit more, less, or will not benefit from any specific treatment after an acute coronary syndrome [56, 57]. However, the lack of empirical data validating the effectiveness of AI-based medications in planned clinical trials is the main obstacle to positive AI deployment. Finally, most research on AI's applications has been led in the business setting; therefore, we need more information on how it affects the patient outcome. Thus far, most of healthcare AI research has been done in non-clinical settings; consequently, generalizing research results might be challenging. Moreover, while AI can be used in basic clinical tasks, it is difficult to see how difficult tasks, like brain surgeries, could be automated: in these tasks, a doctor must quickly change his/her approach once they look into the patient.

5 Future Perspectives: A New Approach to Optimize Healthcare

In conclusion, AI looks well-positioned to revolutionize the healthcare industry. AI-based systems support doctors in daily medical activities, improving disease detection and treatment, nevertheless are unlikely to replace the physician-patient relationship. With this new feature of guided therapy, the doctor can provide immediate and tailored treatment to "treat the patient, not the disease." However, the exploitation of AI in the healthcare field is still in its early phases, and its potential is not entirely explored. Indeed, AI allows the integration of clinical, demographic, biochemical, and instrumental data to develop models that physicians and healthcare providers could use to improve disease management and test new drug therapies using multi-omics methods. Based on the reported data, it has been concluded that AI can help improve healthcare workers' knowledge, enabling them to spend more time in direct patient care and reduce fatigue.

References

1. Witten I, Frank E, Hall M (2011) Data mining: practical machine learning tools and techniques, 3rd edn. Morgan Kaufmann. https://www.sciencedirect.com/science/article/pii/B9780123748560000237
2. Fayyad U, Piatetsky-Shapiro G, Smyth P (1996) Knowledge discovery and data mining: towards a unifying framework. In: Proceedings of the second international conference on knowledge

discovery and data mining, pp 82–88

3. Visco V, Ferruzzi G, Nicastro F, Virtuoso N, Carrizzo A, Galasso G, Vecchione C, Ciccarelli M (2021) Artificial Intelligence as a business partner in cardiovascular precision medicine: an emerging approach for disease detection and treatment optimization. Curr Med Chem 28:6569–6590
4. Johnson K, Torres Soto J, Glicksberg B, Shameer K, Miotto R, Ali M, Ashley E, Dudley J (2018) Artificial intelligence in cardiology. J Am Coll Cardiol 71:2668–2679
5. Ciccarelli M, Giallauria F, Carrizzo A, Visco V, Silverio A, Cesaro A, Calabrò P, De Luca N, Mancusi C, Masarone D, Pacileo G, Tourkmani N, Vigorito C, Vecchione C (2023) Artificial intelligence in cardiovascular prevention: new ways will open new doors. J Cardiovasc Med (Hagerstown) 24:e106–e115
6. Dorado-Diaz P, Sampedro-Gomez J, Vicente-Palacios V, Sánchez P (2019) Applications of artificial intelligence in cardiology. The future is already here. Rev Esp Cardiol (Engl Ed) 72:1065–1075
7. Conroy R, Pyörälä K, Fitzgerald A, Sans S, Menotti A, De Backer G, De Bacquer D, Ducimetière P, Jousilahti P, Keil U, Njølstad I, Oganov R, Thomsen T, Tunstall-Pedoe H, Tverdal A, Wedel H, Whincup P, Wilhelmsen L, Graham I (2003) SCORE project group Estimation of ten-year risk of fatal cardiovascular disease in Europe. The SCORE project. Eur Heart J 24:987–1003
8. Sabovčik F, Ntalianis E, Cauwenberghs N, Kuznetsova T (2022) Improving predictive performance in incident heart failure using machine learning and multi-center data. Front Cardiovasc Med 9:1011071
9. Niel O, Boussard C, Bastard P (2018) Artificial intelligence can predict GFR decline during the course of ADPKD. Am J Kidney Dis 71:911–912
10. Geddes C, Fox J, Allison M, Boulton-Jones J, Simpson K (1998) An artificial neural network can select patients at high risk of developing progressive IgA nephropathy more accurately than experienced nephrologists. Nephrol Dial Transplant 13:67–71
11. Das A, Ben-Menachem T, Cooper G, Chak A, Sivak M, Gonet J, Wong R (2003) Prediction of outcome in acute lower-gastrointestinal haemorrhage based on an artificial neural network: internal and external validation of a predictive model. Lancet 362:1261–1266
12. Sato F, Shimada Y, Selaru F, Shibata D, Maeda M, Watanabe G, Mori Y, Stass S, Imamura M, Meltzer S (2005) Prediction of survival in patients with esophageal carcinoma using artificial neural networks. Cancer 103:1596–1605
13. Shulman E, Kargoli F, Aagaard P, Hoch E, Di Biase L, Fisher J, Gross J, Kim S, Krumerman A, Ferrick K (2016) Validation of the Framingham heart study and CHARGE-AF risk scores for atrial fibrillation in Hispanics, African-Americans, and non-Hispanic whites. Am J Cardiol 117:76–83
14. Pfister R, Brägelmann J, Michels G, Wareham N, Luben R, Khaw K (2015) Performance of the CHARGE-AF risk model for incident atrial fibrillation in the EPIC Norfolk cohort. Eur J Prev Cardiol 22:932–939
15. Ganapathy N, Baumgärtel D, Deserno T (2021) Automatic detection of atrial fibrillation in ECG using co-occurrence patterns of dynamic symbol assignment and machine learning. Sensors (Basel) 21:3542
16. Wang L, Yan Z, Yang Y, Chen J, Yang T, Kuo I, Abu P, Huang P, Chen C, Chen S (2021) A classification and prediction hybrid model construction with the IQPSO-SVM algorithm for atrial fibrillation arrhythmia. Sensors (Basel) 21:5222
17. Parsi A, Glavin M, Jones E, Byrne D (2021) Prediction of paroxysmal atrial fibrillation using new heart rate variability features. Comput Biol Med 133:104367
18. Jo Y, Cho Y, Lee S, Kwon J, Kim K, Jeon K, Cho S, Park J, Oh B (2021) Explainable artificial intelligence to detect atrial fibrillation using electrocardiogram. Int J Cardiol 328:104–110
19. Ribeiro A, Ribeiro M, Paixão G, Oliveira D, Gomes P, Canazart J, Ferreira M, Andersson C, Macfarlane P, Meira W, Schön T, Ribeiro A (2020) Automatic diagnosis of the 12-lead ECG using a deep neural network. Nat Commun 11:1760

20. Eerikainen L, Bonomi A, Schipper F, Dekker L, Morree H, Vullings R, Aarts R (2020) Detecting atrial fibrillation and atrial flutter in daily life using photoplethysmography data. IEEE J Biomed Health Inform 24:1610–1618
21. Kwon S, Hong J, Choi E, Lee B, Baik C, Lee E, Jeong E, Koo B, Oh S, Yi Y (2020) Detection of atrial fibrillation using a ring-type wearable device (CardioTracker) and deep learning analysis of photoplethysmography signals: Prospective observational proof-of-concept study. J Med Internet Res 22:e16443
22. Arnett D, Blumenthal R, Albert M, Buroker A, Goldberger Z, Hahn E, Himmelfarb C, Khera A, Lloyd-Jones D, McEvoy J, Michos E, Miedema M, Muñoz D, Smith S, Virani S, Williams K, Yeboah J, Ziaeian B (2019) 2019 ACC/AHA guideline on the primary prevention of cardiovascular disease: a report of the American college of cardiology/American heart association task force on clinical practice guidelines. Circulation 140:e596–e646
23. Dogan M, Grumbach I, Michaelson J, Philibert R (2018) Integrated genetic and epigenetic prediction of coronary heart disease in the Framingham heart study. PLoS One 13:e0190549
24. D'Ascenzo F, De Filippo O, Gallone G, Mittone G, Deriu M, Iannaccone M, Ariza-Solé A, Liebetrau C, Manzano-Fernández S, Quadri G, Kinnaird T, Campo G, Simao Henriques J, Hughes J, Dominguez-Rodriguez A, Aldinucci M, Morbiducci U, Patti G, Raposeiras-Roubin S, Abu-Assi E, De Ferrari G (2021) PRAISE study group Machine learning-based prediction of adverse events following an acute coronary syndrome (PRAISE): a modelling study of pooled datasets. Lancet 397:199–207
25. Rahmani R, Niazi P, Naseri M, Neishabouri M, Farzanefar S, Eftekhari M, Derakhshan F, Mollazadeh R, Meysami A, Abbasi M (2019) Precisión diagnóstica mejorada para la imagen de perfusión miocárdica usando redes neuronales artificiales en diferentes variables de entrada incluyendo datos clinicos y de cuantificación. Rev Esp Med Nucl Imagen Mol (Engl Ed) 38:275–279
26. Betancur J, Commandeur F, Motlagh M, Sharir T, Einstein A, Bokhari S, Fish M, Ruddy T, Kaufmann P, Sinusas A, Miller E, Bateman T, Dorbala S, Di Carli M, Germano G, Otaki Y, Tamarappoo B, Dey D, Berman D, Slomka P (2018) Deep learning for prediction of obstructive disease from fast myocardial perfusion SPECT: a multicenter study. JACC Cardiovasc Imaging 11:1654–1663
27. Aljaaf A, Al-Jumeily D, Hussain A, Dawson T, Fergus P, Al-Jumaily M (2015) Predicting the likelihood of heart failure with a multi level risk assessment using decision tree. In: 2015 Third international conference on technological advances in electrical, electronics and computer engineering (TAEECE), pp 101–106
28. Kang Y, McHugh M, Chittams J, Bowles K (2016) Utilizing home healthcare electronic health records for telehomecare patients with heart failure: a decision tree approach to detect associations with rehospitalizations. Comput Inform Nurs 34:175–182
29. Turgeman L, May J (2016) A mixed-ensemble model for hospital readmission. Artif Intell Med 72:72–82
30. Koulaouzidis G, Iakovidis D, Clark A (2016) Telemonitoring predicts in advance heart failure admissions. Int J Cardiol 216:78–84
31. Lorenzoni G, Sabato S, Lanera C, Bottigliengo D, Minto C, Ocagli H, De Paolis P, Gregori D, Iliceto S, Pisanò F (2019) Comparison of machine learning techniques for prediction of hospitalization in heart failure patients. J Clin Med 8:1298
32. Jutel A (2009) Sociology of diagnosis: a preliminary review. Sociol Health Illn 31:278–299
33. Holmboe E, Durning S (2014) Assessing clinical reasoning: moving from in vitro to in vivo. Diagnosis (Berlin) 1:111–117
34. Al-Antari M (2023) Artificial intelligence for medical diagnostics-existing and future AI technology!. Diagnostics (Basel) 13
35. Yang Y, Bang C (2019) Application of artificial intelligence in gastroenterology. World J Gastroenterol 25:1666–1683
36. Fernández-Esparrach G, Bernal J, López-Cerón M, Córdova H, Sánchez-Montes C, Miguel C, Sánchez F (2016) Exploring the clinical potential of an automatic colonic polyp detection method based on the creation of energy maps. Endoscopy 48:837–842

37. Pace F, Buscema M, Dominici P, Intraligi M, Baldi F, Cestari R, Passaretti S, Bianchi Porro G, Grossi E (2005) Artificial neural networks are able to recognize gastro-oesophageal reflux disease patients solely on the basis of clinical data. Eur J Gastroenterol Hepatol 17:605–610
38. Lahner E, Grossi E, Intraligi M, Buscema M, Corleto V, Delle Fave G, Annibale B (2005) Possible contribution of artificial neural networks and linear discriminant analysis in recognition of patients with suspected atrophic body gastritis. World J Gastroenterol 11:5867–5873
39. Liu L, Zhao S, Chen H, Wang A (2020) A new machine learning method for identifying Alzheimer's disease. Simul Model Pract Theory 99. https://api.semanticscholar.org/CorpusID:209093045
40. Lodha P, Talele A, Degaonkar K (2018) Diagnosis of alzheimer's disease using machine learning. In: 2018 fourth international conference on computing communication control and automation (ICCUBEA), pp 1–4
41. Arsalan M, Owais M, Mahmood T, Cho S, Park K (2019) Aiding the diagnosis of diabetic and hypertensive retinopathy using artificial intelligence-based semantic segmentation. J Clin Med 8:1446
42. Pizzichetta M, Talamini R, Marghoob A, Soyer H, Argenziano G, Bono R, Corradin M, De Giorgi V, Gonzalez M, Kolm I, Kopf A, Malvehy J, Nami N, Oliviero M, Pellacani G, Puig S, Rabinovitz H, Rubegni P, Seidenari S, Stanganelli I, Veronesi A, Zalaudek I, Zampieri P, Menzies S (2013) Negative pigment network: an additional dermoscopic feature for the diagnosis of melanoma. J Am Acad Dermatol 68:552–559
43. Han S, Kim M, Lim W, Park G, Park I, Chang S (2018) Classification of the clinical images for benign and malignant cutaneous tumors using a deep learning algorithm. J Invest Dermatol 138:1529–1538
44. Navarrete-Dechent C, Dusza S, Liopyris K, Marghoob A, Halpern A, Marchetti M (2018) Automated dermatological diagnosis: hype or reality? J Invest Dermatol 138:2277–2279
45. Buch V, Ahmed I, Maruthappu M (2018) Artificial intelligence in medicine: current trends and future possibilities. Br J Gen Pract 68:143–144
46. Kelly C, Karthikesalingam A, Suleyman M, Corrado G, King D (2019) Key challenges for delivering clinical impact with artificial intelligence. BMC Med 17:195
47. Mathur S, Sutton J (2017) Personalized medicine could transform healthcare. Biomed Rep 7:3–5
48. Sadée W, Dai Z (2005) Pharmacogenetics/genomics and personalized medicine. Hum Mol Genet 14(Spec No. 2):R207-14
49. Mani Kiran S, Nori L (2019) Artificial intelligence: milestones and role in pharma and healthcare sector
50. Surabhi S, Singh B (2018) Computer aided drug design: an overview. J Drug Deliv Ther 8:504–509
51. Lee W, Ahn S, Chung J, Kim K, Kwon K, Kim Y, Sym S, Shin D, Park I, Lee U, Baek J (2018) Assessing concordance with Watson for Oncology, a cognitive computing decision support system for colon cancer treatment in Korea. JCO Clin Cancer Inform 2:1–8
52. Williams B, Mancia G, Spiering W, Agabiti Rosei E, Azizi M, Burnier M, Clement D, Coca A, Simone G, Dominiczak A, Kahan T, Mahfoud F, Redon J, Ruilope L, Zanchetti A, Kerins M, Kjeldsen S, Kreutz R, Laurent S, Lip G, McManus R, Narkiewicz K, Ruschitzka F, Schmieder R, Shlyakhto E, Tsioufis C, Aboyans V, Desormais I (2018) ESC scientific document group. In: 2018 ESC/ESH guidelines for the management of arterial hypertension. Eur Heart J 39:3021–3104
53. Visco V, Izzo C, Mancusi C, Rispoli A, Tedeschi M, Virtuoso N, Giano A, Gioia R, Melfi A, Serio B, Rusciano M, Di Pietro P, Bramanti A, Galasso G, D'Angelo G, Carrizzo A, Vecchione C, Ciccarelli M (2023) Artificial intelligence in hypertension management: an ace up your sleeve. J Cardiovasc Dev Dis 10
54. Ye X, Zeng Q, Facelli J, Brixner D, Conway M, Bray B (2020) Predicting optimal hypertension treatment pathways using recurrent neural networks. Int J Med Inform 139:104122

55. Lofaro D, Groccia M, Guido R, Conforti D, Caroleo S, Fragomeni G (2016) Machine learning approaches for supporting patient-specific cardiac rehabilitation programs. In: 2016 computing in cardiology conference (CinC), pp 149–152. https://api.semanticscholar.org/CorpusID:18235769
56. Xue Y, Hu Z, Jing Y, Wu H, Li X, Wang J, Seybert A, Xie X, Lv Q (2020) Efficacy assessment of ticagrelor versus clopidogrel in Chinese patients with acute coronary syndrome undergoing percutaneous coronary intervention by data mining and machine-learning decision tree approaches. J Clin Pharm Ther 45:1076–1086
57. Wang H, Zu Q, Chen J, Yang Z, Ahmed M (2021) Application of artificial intelligence in acute coronary syndrome: a brief literature review. Adv Ther 38:5078–5086

Healthcare & Cyberspace: Medical Cyber-Physical Systems

Rocco Zaccagnino, Alfonso Guarino, Delfina Malandrino, and Gianluca Zaccagnino

Abstract The increasing availability of data sources and devices globally has elevated the significance of *cyber-physical systems* (CPS) across various domains, notably in Industry 4.0. CPS plays a crucial role by bridging the gap between *physical systems* and *virtual components*. These systems integrate computational and physical processes, monitored by computers and embedded networks, engaging in continuous feedback mechanisms. The central aim of CPS is to establish bidirectional interaction channels between the physical and cyber realms, creating an *Internet of Things* (IoT) within the physical space. This technology is driving significant progress across various domains, with a particular impact on healthcare, forming the basis for medical *Digital Twins* used for monitoring patients' lifestyles, and providing alerts to patients and about prescriptions and medical consultations. To date, a comprehensive examination of the guidelines essential for realizing the full potential of technologies in critical areas such has healthcare has been lacking. The healthcare sector increasingly demands advanced techniques to offer patients optimal assistance and drive digital transformation initiatives forward. This study seeks to elucidate these guidelines, along with outlining integration methods in healthcare.

1 Cyber-Physical Systems Towards Healthcare

The advent of new technological infrastructures, driven by advances in Information and Communication Technology, has significantly impacted various facets of life, including *Healthcare* [1]. In response to these developments, the *World Health*

R. Zaccagnino (✉) · A. Guarino · D. Malandrino
University of Salerno, Fisciano, SA, Italy
e-mail: rzaccagnino@unisa.it

G. Zaccagnino
TopNetwork SPA, Via Simone Martini 143, 00142 Roma, Italy

M. Ficco and G. D'Angelo (eds.), *Artificial Intelligence Techniques for Analysing Sensitive Data in Medical Cyber-Physical Systems*, Engineering Cyber-Physical Systems and Critical Infrastructures 12, https://doi.org/10.1007/978-3-031-70775-9_2

Organization[1] has emphasized the importance of developing "intelligent" infrastructures [2] to provide remote healthcare facilities. This enables patients to monitor their clinical condition without the need to physically visit clinics or hospitals for routine health checks. The scientific community has taken numerous steps to meet these requirements, aiming to enhance the effectiveness of the distribution of remote intelligent real-time health services. In this context, *Cyber-Physical Systems* (CPS) [3, 4] have garnered significant attention due to their scalability and interoperability with emerging technologies. They have become a crucial innovation of Industry 4.0, attracting applications across diverse domains [5].

The role of CPS is to ensure, utilizing various communication technologies [6], a range of real-time services [7] by interconnecting the *physical space*, which includes physical devices responsible for collecting users' parameters (e.g., health information), with the *cyberspace*, also known as the *virtual space*. The cyberspace encompasses *(i)* intelligent embedded devices for decision-making and *(ii)* cutting-edge devices for storing exponentially growing data [8, 9].

The development of a CPS must guarantee some properties, called *CPS features*:

- *highly certain:* ability to manage in real-time highly unpredictable environments,
- *heterogeneity:* ability to manage three levels of heterogeneity, namely the heterogeneity of the devices used at the *perception* level, the heterogeneity of *network communication*, and the heterogeneity of *computational* technologies,
- *interoperability:* multiple levels share information using distinct technologies,
- *flexibility:* interoperability among storage, computing, and network technologies,
- *close integration:* closed integration of cyberspace and physical space, and capability of synchronization among the different components in the two spaces,
- *quick response:* ability to respond in real-time to evolution in the physical space thanks to the faster communication and computing ability of nodes in cyberspace.

1.1 Medical Cyber-Physical Systems

CPS technology is driving significant progress in Healthcare 4.0, representing the fourth revolution in healthcare [10]. This is due to several factors that have opened up extensive opportunities for future healthcare research. These factors encompass the expansion of research in medical devices and sensors, the advent of the era of medical big data, and the ability to integrate CPS with emerging Industry 4.0 technologies like Cloud Computing and advanced Data Mining techniques.

From this investigation, the concept of Medical Cyber-Physical Systems (MCPS) emerged, referring to intelligent systems with the capability to deliver real-time healthcare services [11]. Their objective is to improve patient care by remotely analyzing health parameters using medical devices and advanced sensors, all while

[1] https://www.who.int/.

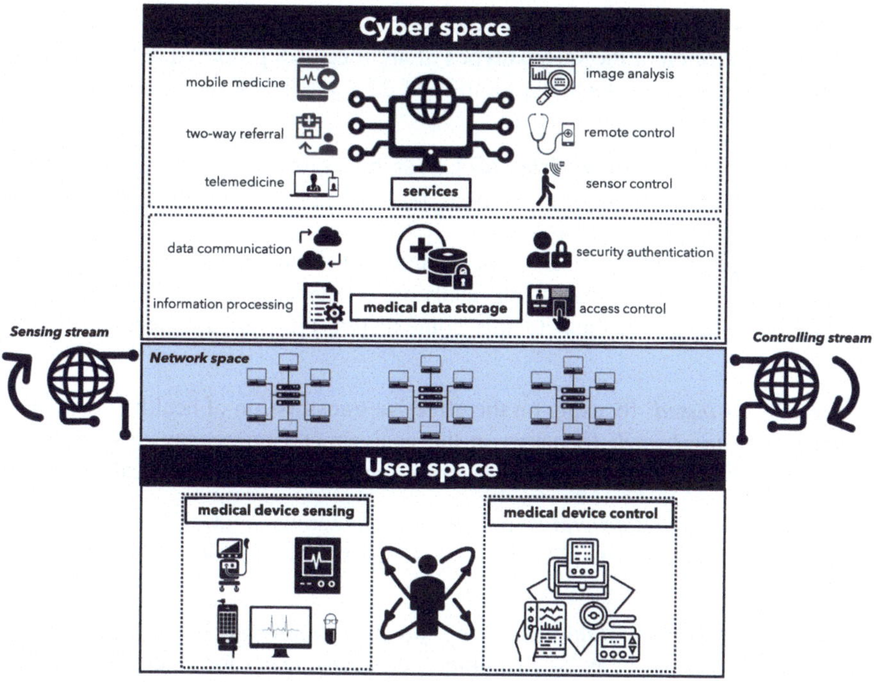

Fig. 1 Framework of Medical Cyber-Physical Systems (MCPS)

maintaining the security and privacy of patient data. MCPS is applicable in the contemporary medical landscape, marked by ongoing interaction between cyberspace (encompassing the network space) and the physical medical space (encompassing the user space), as depicted in Fig. 1

The medical physical space is divided into two main segments: firstly, the *device space* (focused on sensing and controlling), which incorporates diverse real-time healthcare sensing devices, and secondly, the *user space*, encompassing various user types. Its function is to establish a two-way communication link with cyberspace, consistently offering sensing information through *sensing* devices, and reciprocally, receiving information to *control* the physical devices.

The cyberspace serves as the neural center of the MCPS, responsible for processing, storing, and managing health information. It receives sensing information from the medical physical space through the *network space*, which includes network transmission systems. This information is then identified, stored, analyzed, and processed to generate feedback control information, which is subsequently transmitted to the medical physical space. MCPS utilizes dynamic data from the sensor network and monitoring system to construct the medical system through an interaction between cyberspace and physical space. The focus lies on the interaction between the computer system and the physical medical system, ensuring a continuous real-time update of data and feedback.

Below, we will delve into various aspects of MCPS development, beginning with a detailed description of the various types of architectures proposed in the literature and progressing to an in-depth exploration of data storage, security, and protection techniques.

Architecture and Framework The MPCS architectures and frameworks can be analyzed from three main perspectives: *(i) Functional/Behavioral*, *(ii) Hierarchy/Structure*, and *(iii) Communication*.

From a *"Functional/Behavioral"* perspective, the fundamental tasks of an MCPS include the collection, transmission, analysis, processing, storage, and access of patients' personal health data. In this context, two types of architectures can be identified:

- *transmission-based:* focusing on the effective transmission of health data [12].
- *data processing-based:* focusing on the analysis and processing of health data (e.g., aggregation, transformation, filtering mining of massive medical big data).

From a *"Hierarchy/Structure"* perspective, numerous solutions have been proposed. As an example, in [23] the authors proposed the following *three-layer* MCPS:

1. *energy collection and data generation:* involves controlling the actions and gestures of the human body by installing wearable sensors for health monitoring; using a microcontroller and communication technology, the data collected by the sensors is then stored in the sensor nodes.
2. *data preprocessing:* performs data aggregation, transformation, and filtering.
3. *data processing/application:* performs data processing and decision-making.

Mowla et al. [13] proposed the following *four-layer* MCPS architecture:

1. *acquisition layer:* collects human health data from several sensor devices.
2. *preprocessing layer:* stores and processes the data by the previous layer.
3. *cloud layer:* deals with large-scale data processing and computing services.
4. *behavior layer:* provides a health data visualization service used by the medical staff to analyze medical conditions and make informed decisions.

In [12], the authors proposed an MCPS including the following five components:

1. *hybrid sensor:* sensing devices connected/close to the patient (data sources).
2. *patient data aggregator:* a communication hub deployed near patients that transmits the medical data collected by a wireless body area network; this network consists of multiple sensor nodes measuring different vital signs, and the data is transmitted to the infrastructure.
3. *mobile/infrastructure edge node:* handles data management between sources and the Cloud, specifically, it integrates medical and non-medical data from various sources, extracts relevant information, and forwards the processed data.
4. *edge cloud:* utilized for data storage, analysis, and management of human health data, typically associated with a hospital, it plays a role in monitoring/recording the status of patients and providing medical assistance when necessary.

5. *monitoring/service providers:* groups of physical subjects (e.g., doctors, relatives of patients) or intelligent entities (e.g., intelligent ambulances) that can offer prevention, treatment, emergency, or rehabilitation services to patients.

In this category, another intriguing example is presented in [14], where a smart medical framework based on *homomorphic encryption* is detailed. All data tasks, including aggregation, storage, and processing, are carried out in a privacy-preserving manner. The framework comprises four main components:

1. *community members:* healthy people, elderly patients, and hospitalized patients in the smart community whose biological data is analyzed through wired/wireless sensors, encrypted, and then transmitted to the cloud for storage.
2. *IoT gateway:* used for effective local analysis and processing of medical data in the community and local diagnostic feedback.
3. *cloud database*: the database technology used for the storage of the health data received by the community in encrypted form.
4. *anomaly detection system:* deals with the data analysis of encrypted data to identify any anomalies (attacks).

From a *"Communication"* perspective, considering the communication methods, MCPS architectures are typically organized into the following five types [15].

- *client/server:* involving two main types of entities, namely a *client* process that interacts with a single *server* process to access shared resources.
- *point-to-point:* the entities (*processes*) have the same role and interact as *peers*.
- *distributed:* each process consists of two types of entities, with one receiving local/remote calls, while the other only receives local calls.
- *event-based distributed services:* rely on *indirect communication* through a *mediation* mechanism to avoid direct coupling between senders and receivers.

The most commonly used architecture in this category is the *fog-cloud-based layered* architecture, where the physical space and the cyberspace are organized into the following layers:

1. *data collection:* responsible for directly collecting *critical health parameters* from the user's environment; these parameters include respiratory rate, heart rate, heart rate variability, sleep monitoring, electrocardiogram (ECG), and blood sugar levels, data can be collected through various forms, including advanced IoT-based medical sensors, radio frequency identification tags, Wireless Body Area Networks, actuators, and mobile apps, these devices can be placed near users or inside the user's body to detect internal health attributes [16, 17], the collected data is then transmitted to different levels to generate valuable information [18].
2. *data transmission:* an intermediate layer between cyberspace and the data collection layer; it is responsible for transmitting health data, and implementing security mechanisms to ensure the secure dissemination of data in cyberspace.

3. *cyberspace:* collects patient data, extracts useful information, and employs various decision algorithms to make decisions; these decisions are then transmitted through the application layer; it also includes numerous sub-components responsible for managing the data collected from devices in the physical space.
4. *fog:* responsible for the short-term analysis of sensitive health data, the results are transmitted to the respective users and stored in cloud storage in cyberspace for future analysis, in the decision phase, it utilizes various data mining and machine learning (ML) algorithms, one of its most crucial services is the cloud storage of users' information, continuously updated based on their current medical conditions, and providing insight results to be shared with interested parties such as patients, medical professionals, or family members.
5. *application:* deals with providing the interface to transmit/receive information.

Modeling and Verification

An MCPS is utilized in various stages of the medical process, including postoperative rehabilitation, drug administration, and diagnosis. Therefore, an MCPS must adhere to safety requirements set by standards such as *ISO 60601*,[2] which outlines several safety aspects to prevent medical devices from causing harm to patients. In [19], the authors introduced the *CPS-MAS* MCPS modeling and verification framework for security verification. In this method, the operation of an MCPS is simulated as a *state machine*, leading to an abstract model. This model is subsequently analyzed by examining the behavior when the model parameters change. These parameters are then compared with the security requirements expressed by the constraints.

In [20], the authors proposed an end-to-end model-based approach for medical device software development, highlighting various steps, including model creation and verification, code generation, and code execution.

Identification and Sign Sensing

The unique identification of nodes in MCPS is crucial for secure access authentication and communication between nodes. Consequently, the implementation of a secure authentication mechanism to establish effective communication between nodes is an inevitable challenge in the field of MCPS research.

Regarding *"Sign sensing"*, two main types of signs must be considered:

- *vital signs' data:* the literature recognizes *blood pressure*, *breath*, *pulse*, and *body temperature* as the four essential indicators of the human body; in individuals without health issues, these indicators exhibit internal correlations and remain relatively stable within a specified range; conversely, deviations in these indicators can signal the onset of a disease process when the human body undergoes an illness or injury; consequently, the ongoing monitoring and documentation of human vital signs constitute crucial tasks for establishing clinical diagnoses and ensuring timely treatment; MCPS utilizes interactive smart devices to continually and

[2] https://webstore.iec.ch/publication/22261.

efficiently observe alterations in patients' vital signs, enabling medical professionals, patients, or their families to access real-time analyses based on the data gleaned from continuous monitoring.

- *sign sensors:* these sensors play a crucial role in the functioning of an MCPS by collecting vital signs' data from the human body; they can be categorized based on various criteria, including *"human body contact ways"*, *"working principle"*, and *"measured quantity"*; regarding *"human body contact ways"*, they fall into *(i) non-contact*, capable of generating a signal without direct contact with the human body, *(ii) contact*, requiring contact with the human body to generate a signal, and *intrusive* sensors, also known as *implantable* sensors, which can be directly implanted in the human body to collect more accurate vital sign data; regarding the *"working principle"*, various types of physical sign sensors exist, such as strain, piezoelectric, capacitive, thermoelectric, inductive, and photoelectric sensors; concerning the *"measured quantity"*, there are sensors designed to measure different vital signs of the human body, such as heart rate, blood pressure, and body temperature sensors.

Data: Storage, Analysis, and Security/Privacy

In the context of MCPS, four primary *storage technologies* need to be emphasized:

- *database:* extensively employed in the realm of medical data storage; as highlighted in [21], a community health center typically comprises the nursing center workstation, medical personnel, and the database server used for storing personal data and doctor information; a crucial aspect involves managing data related to the utilization and management of medical supplies. To address this challenge, developing a system where a network connects patient data, inventory data, service information, provider information, and doctor information from various sources and stores them in the database might be beneficial [22].
- *big data:* medical big data, consisting of complex high-dimensional data, presents numerous critical challenges; an accurate big data processing technology can significantly save processing time, incorporating advanced data mining techniques like association rule learning and natural language processing; these techniques can enhance diagnostic accuracy, improve treatment outcomes, and reduce costs and waste [23]; various examples in the literature address the management of medical Big Data, such as [24], which focuses on safeguarding doctors' privacy in the context of COVID-19 Big Data; the text classification method, in this case, is based on extracting *word embeddings* from medical documents.
- *cloud Computing:* this technology plays a significant role in storing medical Big Data; it offers an effective computing model that can be applied to a *regional medical data center*, i.e., the cloud computing platform service [25], providing storage solutions for large amounts of medical data; presently, intelligent medical care, based on the edge computing function in cloud computing and next-generation wireless network technology, forms the foundation for techniques used in hospitals to monitor patients' status while providing necessary support [12].

- *blockchain:* medical data possesses unique characteristics that require real-time sharing and protection to ensure patient privacy; blockchain can be utilized to facilitate secure data sharing, overcoming the limitations of traditional access to medical data, which often demand significant resources and time for permission review and data verification, leading to substantial risks and authorizations [26].

Ensuring the integrity, effectiveness, and authenticity of medical data (*security*) and allowing access to patients' medical data only for authorized users (*privacy*) are essential aspects of an MCPS. The main security/privacy requirements are:

1. *integrity:* the data stored in the database must maintain the *correct state* at various levels, including entity integrity, domain integrity, referential integrity, and user-defined integrity.
2. *availability:* authorized users can access data only for which they are authorized.
3. *audit:* database operation control that alerts to risky database behavior and blocks malicious activities.
4. *information privacy:* patient information records can be categorized in: *(i)* general documents that can be accessed publicly, and *(ii)* sensitive records, such as information on the patient's personal identity and medical diagnosis; access to sensitive records necessitates meticulous access control techniques.

2 Conclusions

The escalating ubiquity of data sources and technological devices on a global scale has heightened the significance of cyber-physical systems (CPS), particularly within the context of Industry 4.0. Serving as a crucial nexus between physical systems and virtual elements, CPS seamlessly integrates computational and physical processes, all orchestrated and monitored by computers and embedded networks with continuous feedback loops.

In the realm of healthcare, CPS plays a pivotal role, enabling real-time monitoring of patients' lifestyles, dietary habits, and health parameters. This advanced technological infrastructure not only provides alerts to patients regarding prescriptions and medical consultations but also facilitates the selection of optimal medications. Furthermore, it contributes to predicting the outcomes of specific surgical procedures, showcasing its transformative impact on healthcare delivery.

Despite the evident advantages, there has been a notable absence of a comprehensive examination of guidelines essential for harnessing the full potential of CPS in critical areas such as healthcare. The healthcare sector is increasingly recognizing the need for advanced techniques to provide patients with optimal assistance and to drive forward digital transformation initiatives.

This study aims to address this gap by elucidating and organizing guidelines, and highlighting significant implications for healthcare.

References

1. Abujassar RS (2020) A new smart algorithm to provide an integrity path in networks for improving the abroad healthcare surgery setting. Int J Innov Comput Inf Control 16(3):1047–1058
2. W.H.O. Assembly (2005) World health organization assembly. https://www.who.int/healthacademy/media/WHA58-28-en.pdf?ua=1. Accessed Feb 2021
3. Lee J, Bagheri B, Kao H-A (2015) A cyber-physical systems architecture for industry 4.0-based manufacturing systems. Manuf Lett 3:18–23
4. Gupta V, Singh Gill H, Singh P, Kaur R (2018) An energy efficient fog-cloud based architecture for healthcare. J Stat Manag Syst 21(4):529–37
5. Aceto G, Persico V, Pescapé A (2020) Industry 4.0 and health: internet of things, big data, and cloud computing for healthcare 4.0. J Ind Inf Integr 18:100129
6. Rao CS, Shankar RS, Murthy K (2019) Cyber-physical systems-an overview: smart intelligent computing and applications, p 489–97
7. Humayed A, Lin J, Li F, Luo B (2017) Cyber-physical systems security-a survey. IEEE Internet Things J 4(6):1802–31
8. Munir A, Kansakar P, Khan SU (2017) IFCIoT: integrated fog cloud IoT: a novel architectural paradigm for the future internet of things. IEEE Consum Electron Mag 6(3):74–82
9. Chaâri R, Ellouze F, Koubâa A, Qureshi B, Pereira N, Youssef H, Tovar E (2016) Cyber-physical systems clouds: a survey. Comput Netw 108:260–78
10. Tortorella GL, Fogliatto FS, Mac Cawley Vergara A, Vassolo R, Sawhney R (2020) Healthcare 4.0: trends, challenges and research directions. Prod Plan Control 31(15):1245–60. 23
11. Zhang Y, Qiu M, Tsai C-W, Hassan MM, Alamri A (2015) Health-CPS: healthcare cyber-physical system assisted by cloud and big data. IEEE Syst J 11(1):88–95
12. Abdellatif AA, Mohamed A, Chiasserini CF, Tlili M, Erbad A (2019) Edge computing for smart health: context-aware approaches, opportunities, and challenges. IEEE Netw 33(3):196–203
13. Mowla N, Doh I, Chae K (2018) Selective fuzzy ensemble learner for cognitive detection of bio-identifiable modality spoofing in MCPS. In: Proceedings of 20th international conference on advanced communication and technology (ICACT), pp 63–67
14. Alabdulatif A, Khalil I, Yi X, Guizani M (2019) Secure edge of things for smart healthcare surveillance framework. IEEE Access 7:31010–31021
15. Roehrs A, da Costa CA, da Rosa Righi R, da Silva VF, Goldim JR, Schmidt DC (2019) Analyzing the performance of a blockchain-based personal health record implementation. J Biomed Inf 92. Art. no. 103140
16. Verma P, Sood SK (2018) Fog assisted-IoT enabled patient health monitoring in smart homes. IEEE Internet Things J 5(3):1789–96
17. Raghupathi W, Raghupathi V (2014) Big data analytics in healthcare: promise and potential. Health Inf Sci Syst. 2(1):1–10
18. Anuradha J et al (2015) A brief introduction on big data 5 vs characteristics and Hadoop technology. Proc Comput Sci. 48:319–24
19. Banerjee A, Venkatasubramanian KK, Mukherjee T, Gupta SKS (2011) Ensuring safety, security, and sustainability of mission-critical cyber-physical systems. Proc IEEE 100(1):283–299
20. Murugesan A et al (2014) From requirements to code: Model based development of a medical cyber physical system. In: Proceedings of international workshop software engineering health care, pp 96–112
21. Wu Z et al (2016) A community health service architecture based on the Internet of Things on health-care. Chin J Biomed Eng 25(3):114–120
22. Xu E, Wermus M, Bauman DB (2011) Development of an integrated medical supply information system. Enterprise Inf Syst 5(3):385–399
23. Din S, Paul A (2019) Smart health monitoring and management system: Toward autonomous wearable sensing for Internet of Things using big data analytics. Future Gener Comput Syst 91:611–619

24. Ma B et al (2020) Assuring privacy-preservation in mining medical text materials for COVID-19 cases-A natural language processing perspective. In: Proceedings of VLDB international workshop Very Large Internet Things (VLIoT), pp 6–13
25. Jiemin Z, Jinsheng L (2010) The district medical data center based on cloud computing. In: Proceedings of 5th international conference on computer science education, pp 1424–1427
26. Xue T et al (2017) Research on medical data sharing model based on blockchain. Acta Automatica Sinica 43(9):1555–1562
27. Grieves M (2014) Digital twin: manufacturing excellence through virtual factory replication. White paper 1(2014):1–7
28. Liu Y, Zhang L, Yang Y, Zhou L, Ren L, Wang F, Deen MJ (2019) A novel cloudbased framework for the elderly healthcare services using digital twin. IEEE Access 7:49088–49101
29. Elayan H, Aloqaily M, Guizani M (2021) Digital twin for intelligent context-aware IoT healthcare systems. IEEE Internet Things J 8(23):16749–16757
30. Erol T, Mendi AF, Dogan D (2020) The digital twin revolution in healthcare. In: 2020 4th international symposium on multidisciplinary studies and innovative technologies (ISMSIT). IEEE, pp 1–7
31. Haleem A, Javaid M, Singh RP, Suman R (2022) Medical 4.0 Technologies for Healthcare: Features, Capabilities, and Applications, Internet of Things and Cyber-Physical Systems
32. Feng Y, Chen X, Zhao J (2018) Create the individualised digital twin for noninvasive precise pulmonary healthcare. Significances Bioeng Biosci 1(2)
33. Fuller A, Fan Z, Day C, Barlow C (2020) Digital twin: enabling technologies, challenges and open research. IEEE Access 8:108952–108971
34. Rivera LF, Jimenez M, Angara P, Villegas NM, Tamura G, Müller GA (2019) Towards continuous monitoring in personalised healthcare through digital twins. In: Proceedings of the 29th annual international conference on computer science and software engineering, pp 329–335
35. Ahmed H, Devoto L (2021) The potential of a digital twin in surgery. Surg Innovat 28(4):509–510
36. Javaid M, Haleem A, Singh RP, Khan S, Suman R (2021) Blockchain technology applications for industry 4.0: a literature-based review, blockchain: research and applications, 100027
37. Barricelli BR, Casiraghi E, Fogli D (2019) A survey on digital twin: definitions, characteristics, applications, and design implications. IEEE Access 7:167653–167671
38. Popa EO, van Hilten M, Oosterkamp E, Bogaardt MJ (2021) The use of digital twins in healthcare: socio-ethical benefits and socio-ethical risks. Life Sci Soc Policy 17(1):1–25
39. De Benedictis A, Mazzocca N, Somma A, Strigaro C (2022) Digital twins in healthcare: an architectural proposal and its application in a social distancing case study. IEEE J Biomed Health Inform. https://doi.org/10.1109/JBHI.2022.3205506
40. Okegbile SD, Cai J, Yi C, Niyato D (2022) Human digital twin for personalised healthcare: vision, architecture and future directions. IEEE Network
41. El Marai O, Taleb T, Song J (2020) Roads infrastructure digital twin: a step toward smarter cities realisation. IEEE Network 35(2):136–143
42. Heininger R, Stary C (2021) Capturing autonomy in its multiple facets: a digital twin approach. In: Proceedings of the 2021 ACM workshop on secure and trustworthy cyber-physical systems, pp 3–12
43. Liu YK, Ong SK, Nee AYC (2022) State-of-the-art survey on digital twin implementations. Adv Manuf 10(1):1–23
44. Madubuike OC, Anumba CJ (2021) Digital twin application in healthcare facilities management. Comput Civil Eng 366–373
45. Song Y, Li Y (2022) Digital twin aided healthcare facility management: a case study of Shanghai Tongji hospital. Construction Research Congress, pp 1145–1155
46. Hu L, Xie N, Kuang Z, Zhao K (2012) Review of cyber-physical system architecture. In: Proceedings -, (2012) 15th ieee international symposium on object/component/service-oriented real-time distributed computing workshops. ISORCW, pp 25–30
47. Liu Y, Peng Y, Wang B, Yao S, Liu Z (2017) Review on cyber-physical systems. IEEE/CAA J Autom Sin 4(1):27–40

48. Win SY, Thianwiboon M (2021) Parametric optimization of NACA 4412 airfoil in ground effect using full factorial design of experiment. Eng J 25(12):9–19
49. Parnianifard A, Rezaie V, Chaudhary S, Imran MA, Wuttisittikulkij L (2021) New adaptive surrogate-based approach combined swarm optimizer assisted less tuning cost of dynamic production- inventory control system. IEEE Access 9:144054–144066
50. Amaruchkul K, Pongsathornwiwat A, Bantadtiang P (2022) Constrained joint replenishment problem with refrigerated vehicles. Eng J 26(1):75–91
51. Wang H, Deng X, Tian F (2012) WiP abstract: a human-centered cyber-physical systematic approach for post-stroke monitoring. In: 2012 IEEE/ACM third international conference on cyber-physical systems, p 209
52. Ghorbani M, Bogdan P (2013) A cyber-physical system approach to artificial pancreas design. In: 2013 international conference on hardware/software codesign and system synthesis, CODES+ISSS, pp 1–10
53. Bécue A, Fourastier Y, Praça I, Savarit A, Baron C, Gradussofs B et al (2018) Cyberfactory#1 - securing the industry 4.0 with cyber-ranges and digital twins. In 2018 14th IEEE international workshop on factory communication systems (WFCS), pp 1–4
54. Bitton R, Gluck T, Stan O, Inokuchi M, Ohta Y, Yamada Y et al (2018) Deriving a cost-effective digital twin of an ICS to facilitate security evaluation. In: Lopez J, Zhou J, Soriano M (eds) Computer security. Springer, Cham, pp 533–554
55. Eckhart M, Ekelhart A (2018) Towards security-aware virtual environments for digital twins. In: Proceedings of the 4th ACM workshop on cyber-physical system security, CPSS '18. ACM, New York, NY, pp 61–72
56. Tauber M, Schmittner C (2018) Enabling security and safety evaluation in industry 4.0 use cases with digital twins. ERCIM News 2018(115)
57. Axelrod CW (2013) Managing the risks of cyber-physical systems. In: (2013) IEEE long island systems. Applications and technology conference (LISAT), pp 1–6
58. Venkatasubramanian KK, Nabar S, Gupta SKS, Poovendran R (2012) Cyber physical security solutions for pervasive health monitoring systems, in E-Healthcare Systems and Wireless Communications: Current and Future Challenges. IGI Global, pp 143–162
59. Arney D, Pajic M, Goldman JM, Lee I, Mangharam R, Sokolsky O (2010) Toward patient safety in closed-loop medical device systems. In: Proceedings of the 1st ACM/IEEE international conference on cyber-physical systems, pp 139–148
60. Barbiero P, Lió P (2020) The computational patient has diabetes and a covid. arXiv:2006.06435
61. Barbiero P, Vinas Torne R, Lió P (2021) Graph representation forecasting of patient's medical conditions: toward a digital twin. Front Genet 12:652907
62. Zitnik M, Agrawal M, Leskovec J (2018) Modeling polypharmacy side effects with graph convolutional networks. Bioinformatics 34:i457–i466. https://doi.org/10.1093/bioinformatics/bty294
63. Ginsburg GS, Willard HF (2009) Genomic and personalized medicine: foundations and applications. Transl Res 154:277–287. https://doi.org/10.1016/j.trsl.2009.09.005.
64. Naylor S, Chen JY (2010) Unraveling human complexity and disease with systems biology and personalized medicine. Pers Med 7:275–289. https://doi.org/10.2217/pme.10.16
65. Laubenbacher R, Sluka JP, Glazier JA (2021) Using digital twins in viral infection. 371:1105–1106. https://doi.org/10.1126/science.abSciencef3370
66. Scarselli F, Gori M, Tsoi AC, Hagenbuchner M, Monfardini G (2008) The graph neural network model. IEEE Trans Neural Netw 20:61–80. https://doi.org/10.1109/TNN.2008.2005605
67. Goodfellow I, Pouget-Abadie J, Mirza M, Xu B, Warde-Farley D, Ozair S et al (2014) Generative adversarial nets. In: Advances in neural information systems. Processing systems, pp 2672–2680
68. Yang J, Li Y, Liu Q, Li L, Feng A, Wang T, Lyu J (2020) Brief introduction of medical database and data mining technology in big data era. J Evid Based Med 13(1):57–69
69. Emilsson V, Thorleifsson G, Zhang B, Leonardson AS, Zink F, Zhu J et al (2008) Genetics of gene expression and its effect on disease. Nature 452:423–428. https://doi.org/10.1038/nature06758

70. Cookson W, Liang L, Abecasis G, Moffatt M, Lathrop M (2009) Mapping complex disease traits with global gene expression. Nat Rev Genet 10:184–194. https://doi.org/10.1038/nrg2537
71. Gamazon ER, Segrè AV, van de Bunt M, Wen X, Xi HS, Hormozdiari F et al (2018) Using an atlas of gene regulation across 44 human tissues to inform complex disease-and trait-associated variation. Nat Genet 50:956–967. https://doi.org/10.1038/s41588-018-0154-4
72. Zaccagnino R, Cirillo A, Guarino A, Lettieri N, Malandrino D, Zaccagnino G (2023) Towards a geometric deep learning-based cyber security: network system intrusion detection using graph neural networks. SECRYPT, pp 394–401
73. Arjovsky M, Chintala S, Bottou L (2017) Wasserstein GAN. arXiv:1701.07875
74. Gulrajani I, Ahmed F, Arjovsky M, Dumoulin V, Courville A (2017) Improved training of Wasserstein GANs. arXiv: 1704.00028
75. Fountain JH, Lappin SL (2019) Physiology, renin angiotensin system. In: StatPearls. StatPearls Publishing
76. Biolchini J, Mian P, Natali A, Travassos G (2005) Systematic review in software engineering. System Engineering And Computer Science Department COPPE/UFRJ, Technical Report ES. 679:45
77. Kitchenham B, Charters S, Others (2007) Guidelines for performing systematic literature reviews in software engineering. UK
78. Roehrs A, Da Costa C, Rosa Righi R, De Oliveira K, Others (2017) Personal health records: a systematic literature review. J Med Internet Res 19:e5876
79. Petticrew M, Roberts H (2008) Systematic reviews in the social sciences: a practical guide. Wiley
80. Aria M, Cuccurullo C (2017) bibliometrix: an r-tool for comprehensive science mapping analysis. J Informet 11:959–975
81. Gobbo E, Guarino A, Cafarelli B, Grilli L, Limone P (2023) Automatic evaluation of open-ended questions for online learning. A systematic mapping. Stud Edu Eval 77:101258
82. Cobo M, López-Herrera A, Herrera-Viedma E, Herrera F (2011) An approach for detecting, quantifying, and visualizing the evolution of a research field: A practical application to the Fuzzy Sets Theory field. J Informet 5:146–166
83. Callon M, Courtial J, Laville F (1991) Co-word analysis as a tool for describing the network of interactions between basic and technological research: the case of polymer chemistry. Scientometrics 22:155–205
84. Aria M, Cuccurullo C, D'Aniello L, Misuraca M, Spano M (2022) Thematic analysis as a new culturomic tool: the social media coverage on COVID-19 pandemic in Italy. Sustainability 14:3643
85. Garg H, Sharma B, Shekhar S, Agarwal R (2022) Spoofing detection system for e-health digital twin using EfficientNet Convolution Neural Network. Multimed Tools Appl 81:26873–26888

Privacy Risk Analysis in Medical Cyber-Physical Systems

Michele Mastroianni

Abstract The healthcare system is experiencing substantial transformation, with information and communication technologies (ICTs) playing a primary role in improving healthcare delivery. Moreover, the global acceptance of Medical Cyber-Physical Systems (M-CPS) has been boosted by the COVID-19 pandemic. Historically, healthcare services functioned autonomously without any form of digital interconnection. M-CPS combines distributed systems to concurrently monitor and control several physiological characteristics of patients. Those new systems incorporates embedded software for controlling medical devices, networking capabilities, and managing intricate physical dynamics through advanced healthcare systems. In this chapter have been described the most renowned methodologies of risk assessment used by academic and security experts, such as Risk Matrix and FMEA, to deal with security and privacy issues, and the common threats and vulnerabilities of cyber-physical systems.

1 Introduction

The healthcare system is currently undergoing a significant change, and the utilization of ICTs has played a crucial role in enhancing healthcare delivery systems. In the healthcare sector, the Covid19 pandemic has led to a greater utilization of Medical Cyber-Physical Systems (M-CPS) in both developed and developing countries. Traditionally, healthcare service delivery systems operated independently, without any cyber connectivity, and were created, certified, and used to treat patients separately. M-CPS has implemented distributed systems specifically designed to simultaneously monitor and regulate several physiological features of the patient. The system consists of embedded software that controls medical devices, incorporates networking capabilities, and manages complex physical dynamics using state-of-the-art healthcare systems.

M. Mastroianni (✉)
Università degli Studi di Foggia, Foggia, Italy
e-mail: michele.mastroianni@unifg.it

M. Ficco and G. D'Angelo (eds.), *Artificial Intelligence Techniques for Analysing Sensitive Data in Medical Cyber-Physical Systems*, Engineering Cyber-Physical Systems and Critical Infrastructures 12, https://doi.org/10.1007/978-3-031-70775-9_3

Within the context of M-CPS, the integration of active user input, such as a smart feedback system, together with digital recordings of patient data, and passive user input, such as biosensors and/or smart devices in healthcare environments, can facilitate the collection of data for effective decision making.

The exponential expansion in the last years of M-CPS applications enables several new issues about security. As wireless technologies become more often utilized for collecting and transmitting data, as well as controlling commands, there is an increasing demand for the development of information security systems in the sector [1]. Moreover, the remote operation of CPS devices and their self-sufficiency give rise to the vulnerability of unauthorized access and breach attempts. Managing a substantial number of devices can lead to the compromise of certain devices. CPS security presents several novel issues:

- identify DDoS attacks in this new environment;
- model security threats;
- assessing vulnerabilities;
- design resilient architectures;
- identify appropriate methodologies for risk management/assessment.

In the last years, the introduction of General Data Processing Regulation (GDPR) in EU countries has led to new constraint in terms of personal data protection, and the health sector is particularly affected with those new issues; moreover, many other countries have approved new regulations in the direction of new limitation to personal data processing [2].

Due to the novelty of the field, studies and research on risk analysis methodologies and frameworks for cyber-physical systems in healthcare is in the early stages. In fact, according to a recent review paper [3], only the 2% of research papers is devoted to the study of security and privacy on health application of Cyber-Physical Systems, and none of them deals with risk analysis. That said, in this chapter have been described the most renowned methodologies of risk assessment used by academic and security experts to deal with security and privacy issues, and the common threats and vulnerabilities of cyber-physical systems.

2 Risk Analysis and Risk Assessment

ISO/IEC 27001:2013 defines information security as the act of safeguarding the confidentiality, integrity, and availability of information. Confidentiality refers to the quality of ensuring that information is not shared or revealed to individuals, entities, or procedures that do not have the required authorization. Integrity refers to the quality of being both accurate and complete. Availability refers to the characteristic of being readily accessible and useable when requested by an authorized individual or institution. Additional attributes, such as genuineness, responsibility, inability to deny, and dependability, may also be implicated.

Table 1 The ISO 31000:2015 PDCA framework

PDCA phase	ELEMENTS
Plan	Context determination Risk Assessment Risk mitigation Residual risk evaluation
Do	Plan implementation
Check	Continuous monitoring and inspection (surveillance)
Act	Risk management maintenance and improvement

To better understand the concept of risk assessment and its relevance, it's crucial to distinguish between governance and risk assessment. This distinction is underscored by its regulation in internationally recognized documents like the ISO 31000:2015 standard for Risk Management. Serving as a potential, principled, though non-binding framework, this standard adopts the PDCA (Plan, Do, Check, Act) cycle, delineated in Table 1. Notably, the initial step, fundamental to risk management, involves identifying and assessing risks.

ISO 31000:2015 defines risk management as a broader concept encompassing estimation, incorporating context establishment and risk actions, namely risk treatment. Within risk assessment, there are three key components: risk *identification*, *analysis*, and *evaluation*.

Risk identification aims to compile lists of risk sources, causes, events potentially impacting objectives, and scenario development. The aforementioned standard outlines various risk types, including those related to business, occupational safety, natural disasters, legalities, illegal activities, fire, and standards compliance. Common techniques for risk identification include surveys, interviews, checklists, experience-based judgments, scenario analysis, and engineering system analysis.

Risk analysis serves as input for risk evaluation and decisions regarding risk treatment. Its procedure involves describing identified risks, grouping related sources, analyzing individual causes' influence, evaluating likelihood and consequences, quantifying risk valorization, identifying influencing factors, prioritizing risks, proposing treatment methods, and defining monitoring measures.

Accordingly, risk assessment holds primary importance in the process of risk evaluation, with estimation serving as an additional aspect. This is because determining the value of a risk stems directly from the analysis conducted, forming the basis for all subsequent methods aimed at assessing risk. The techniques utilized in risk assessment can be categorized into three main groups: qualitative, semi-quantitative (or a blend of qualitative and quantitative), and quantitative.

Qualitative and semi-quantitative risk analysis methods encompass a range of techniques, including polling, SWOT analysis, causal diagrams, expert opinion methods, the Delphi method, preliminary hazard analysis, fault tree/failure mode analysis, event tree analysis, and the use of probability matrices.

Quantitative risk analysis methods include probability theory, mathematical statistics, operational research, sensitivity analysis, scenario methods, error log analysis, event tree analysis, the Monte Carlo method, and modeling and simulation techniques. Notably, the probability and consequence matrix or matrix-based methods are considered foundational for all the aforementioned qualitative risk assessment techniques (Kovacevic et al. [7]).

Risk evaluation involves comparing the level of risk identified during the risk analysis phase against predefined risk criteria established in the risk management context. It entails determining the significance of the risk and deciding on appropriate risk management strategies. If the estimated risk aligns with the established criteria, it is deemed acceptable and may not require additional control measures. However, if it surpasses the criteria, it becomes necessary to prioritize risks and develop strategies to address them. Standards such as ISO-IEC 31010 provide specific guidelines on risk assessment techniques and the regulation of value at risk.

3 The Perspective of the Analysis

The analysis of risks, and consequently the related risk assessment model, can be approached from three perspectives: Asset-driven, Service-driven, and Business-driven (Fig. 1). The primary objective of this classification is to select a specific level of the organization's resources, such as assets, services, or business processes, in order to assess the associated risks. The asset-driven approach focuses on identifying the assets and their corresponding risks at the individual asset level, while the Business-driven perspective identifies threats that directly impact the business processes. The Service-driven view, in contrast, utilizes services as the input for risk assessment and focuses on assessing threats to services rather than assets or business processes. This section provides a comprehensive explanation of the three risk assessment perspectives.

3.1 Asset-Driven Approach

The asset-driven approach is the prevailing stance when it comes to risk assessment. From this point of view, assets are recognized and assessed, followed by the computation of the corresponding risks. In order to attain a specific level of security and/or privacy, the assessment strategy must identify the organization's significant information assets and rigorously assess the risks they are subjected to [4]. The asset-driven method involves identifying the threats and vulnerabilities associated with each asset and developing risk scenarios accordingly. In a medium to large business, there is a vast number of assets. When this number is multiplied by the average number of risk scenarios for each information asset, it results in extensive lists of risk scenarios. These lists must be regularly examined and updated. The primary drawback of the asset-driven perspective is that it renders risk assessment a laborious and prone to

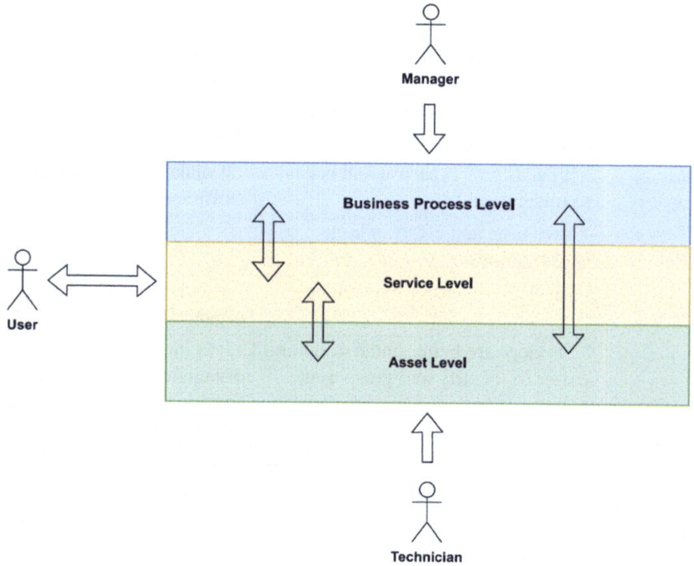

Fig. 1 Relationship between assets, services and business processes

mistakes endeavor. Asset-driven techniques offer a significant benefit in terms of their simplicity and general acceptance, resulting in a plethora of solutions specifically tailored to this perspective.

3.2 Service-Oriented Approach

From a service-oriented standpoint, risks are identified and evaluated according to their effect on the services. From this standpoint, the initial step involves identifying all the crucial services and service packages offered by the firm. Unlike asset-driven techniques, a service-driven viewpoint focuses on identifying risks and weaknesses for each service rather than individual assets. Due to the disparity in quantity between the services and assets of a business, adopting a Service-driven approach is more manageable.

3.3 Business-Driven (or Process-Driven) Approach

The methodologies for conducting security risk assessments in a corporate context are centered around the objectives and the operational procedures that contribute to achieving these objectives [5]. From this standpoint, values are not assigned to assets,

Table 2 The three perspectives of risk assessment (adapted from [6])

	Advantages	Disadvantages
Asset-dr.	- The majority of tools available on the market are designed based on this perspective	- Makes the calculation error-prone for the large number of resources in a medium to large organization
	- It needs less expertise and is easy to understand	- It makes the risk reduction inefficient
	- Save time and effort in large organizations	
Service-dr.	- Services are easier to evaluate	- May be too challenging for small organizations
	- Services are better understood and easier to identify and grasp than assets	- It is more suitable for those organizations, which have a service-based approach in delivering their business processes
Business-dr.	Those risks that have high impact in organization can be seen well in output of risk assessment	- May be too challenging for small organizations
	- Business processes can be better evaluated even based on their financial impact	- It needs more insight into the business and its processes
	- Business processes are better understood and easier to explain for the top management and acquire the needed support	
	- It leads to an efficient risk reduction	

but to processes, that are directly connected to corporate objectives. Understanding and estimating the business operations of a company is simpler and more straightforward compared to information assets. The primary emphasis in the Business-driven viewpoint is on identifying and assessing the business processes and their associated risks and threats. Due to the clear correlation between this viewpoint and the organization's aims and objectives, the risk assessment findings may be more easily elucidated to senior management and obtain the necessary support.

A summary of advantages and disadvantages of the three different approaches is shown in Table 2.

4 Risk Matrix

The main foundation of risk assessment qualitative approaches is the personal experiences and opinions of the risk assessors (members of the risk assessment team), and/or the utilization of existing qualitative data. This method doesn't require knowledge of past dangers, hazards, causes, and consequences, but it does result in a descriptive description of the qualitative risk size (high risk, moderate risk, etc.) at the end of the risk assessment process.

Qualitative criteria indicate the likelihood of undesirable events using terms like "rarely," "amazing," "possible," "probable," or "almost certain," and the size of a damage-consequence using terms like "fatal," "serious," "small," or "negligible." The most widely used subjective criteria in risk assessment qualitative approaches are those measured using qualitative scales [7].

So, risk assessment is subjective by nature, and mistakes can be made in it. The best application of qualitative scales in practice involves three to seven qualitative descriptions, necessitating a strong professional attitude to possibility analysis of dangers and/or threats. More than seven qualitative descriptions can make a method very difficult and subjective to carry out because it makes it difficult for the members of the risk assessment team to fairly accurately identify the qualitative description of the risk factors or constituents.

In qualitative approach to risk assessment, the most well-known techniques is the *Risk Matrix*. In a risk matrix operation, risk assessors (domain experts) are frequently employed to create a logical link between the outcome and the likelihood of the risk assessment of recognized dangers or harmfulness. Additionally, they are applied in accordance with the uniform method's definition to determine the degree or amount of each expert's expected hazards.

The following three steps are used to create a risk matrix: (Step 1) ranks of ordinates are applied to the probability; (Step 2) abscissas are applied to the rank/severity result. When the aforementioned ranking levels are combined, the Risk ranking (Step 3), as depicted in Fig. 2. Information gathering, the initial stage in any risk assessment process, is required to obtain these data (probability and effects).

Experience of academic and practitioners in the real world has demonstrated that "checklists" are a great way to gather data that will help identify potential risks and hazards in the workplace [6]. Examining all parties involved in the work process-administrative and executive bodies as well as end users/workers-is essential to obtaining a complete picture of all potential risks and hazards and, ultimately, a reliable risk assessment procedure. In the most cases, 3×3, 4×4 and 5×5 matrices have been used.

In the Guidance on Risk Assessment redacted by *European Agency for Safety and Health at Work* recommends a 3×3 matrix, which was first defined in the standard OHSAS 18001 and which is shown in Fig. 2. Three levels for the qualitative description of probability (rare, possible and almost certain, as well as consequences (minor, major and serious). Risk is also ascribed three levels, marked as a qualitative description of: low, moderate, and high. The cell of the matrix are colored in green, yellow and red according with the three level of risk.

Usually, the high risk level (red area) represents a risk that must be mitigated, while the yellow area (moderate level of risk) is and area in which the management of the company must choose if mitigate or accept the risk. Green area represents the risk that may be not reduced.

		Result of a dangerous event		
		Minor (1)	Moderate (2)	Serious (3)
Probability of dangerous event	Rare (1)	Low Risk (1)	Low Risk (2)	Moderate Risk (3)
	Possible (2)	Low Risk (2)	Moderate Risk (4)	High Risk (6)
	Almost certain (3)	Moderate Risk (3)	High Risk (6)	High Risk (9)

Fig. 2 An example 3 × 3 risk matrix

5 Failure Mode and Effect Analysis (FMEA) Methodology

Failure Modes and Effects Analysis (FMEA) is an effective tool initially designed to strategically design and implement preventive maintenance systems. The technique was developed by reliability engineers in the 1950s to evaluate the safety of military systems. Subsequently, the utilization of this technique rapidly proliferated in the aerospace and automotive sectors. The effectiveness of this approach has resulted in healthcare centers adopting it to enhance patient safety and their emergency medical services system. This method is commonly employed in the electronics, chemicals, and other production industries to detect, prioritize, and resolve failures, defects, and possible problems [8].

FMEA is a highly effective tool used to identify potential failure modes and their corresponding impacts. Its purpose is to enhance the dependability and safety of complex systems by collecting the required data to make informed decisions regarding risk management. The purpose of this technique is to identify failure modes, their impacts, and remedial actions in order to remove or reduce the likelihood of failure through redesign. Additionally, it aims to establish an efficient maintenance system to minimize the occurrence of probable situations. This strategy has been demonstrated to be one of the most crucial main prevention methods in the design of systems, products, processes, or services. The method is considered one of the most essential risk analysis tools due to its feasibility and application. However, the extensive use of this method in several domains has exposed its faults and limits, so researchers have made significant efforts to improve and enhance it [9].

The approach of this methodology is qualitative, and relies on the expertise of a team of experts. The aim of team building in FMEA is to leverage the unique perspectives and specialized knowledge of individuals. Given that each process focuses on certain features of a product, FMEA teams are established as needed and disbanded once the FMEA project is completed. Since each FMEA team has a distinct objective and duty, there is no necessity to have a permanent FMEA team. In the opinion of the majority of experts, the optimal size for an FMEA team is between 4 and 6 individuals [10]. For a complete description of the methodology, refer to

FMEA Handbook Version 4.2, available for download in the Ford Supplier Portal.[1] A comprehensive review on state-of-art of scientific research about FMEA is [11].

5.1 FMEA Process

Firstly, it is necessary to define the term *Failure* and the related concepts. According to IEC 60812:

- **Failure cause** what led to the item's malfunction
- **Failure mode** refers to the way in which an item fails
- **Failure effect** refers to the consequence of a failure mode in terms of how it affects the operation, function, or status of the item.
- **Failure severity** is the grading or significance of the effect that a failure mode has on the item's operation, its surroundings, or the person operating it.
- **Failure criticality** is a measure of the need to address and mitigate a failure, which takes into account both the severity of the effect and the frequency of its occurrence or other attributes of the failure.

The Failure Mode and Effects Analysis evaluation is based on three parameters: (i) the **Severity**, that is an esteem of the effect of a feared event; (ii) the **Occurrence**, i.e. the probability of occurrence of a feared event; (iii) the probability of **Detection** of the event feared. The total risk assessed in this methodology is called Risk Priority Number (RPN) and is the product of those three factors, as in Formula 1

$$RPN = SxOxD \tag{1}$$

The values for S, O and D are assigned by experts' teams in a qualitative scale. As an example, Christopher [12] uses a 1–5 scale for all the three parameters, as shown in Table 3.

The assessment process is typically employed in the first phases of system development and it encompasses the subsequent stages:

1. List the relevant components of the system.
2. Determine all potential failure modes of the component or system. The team engage in a collaborative brainstorming session to identify the potential ways in which failure can occur. The causes and conditions that are likely to lead to each failure scenario are also provided.
3. Determine the probability (likelihood) of detection **D** and occurrence **O**) of each failure mode.
4. Analyze and evaluate the potential consequences of each failure mode on the system's performance, safety, and the environment.

[1] https://fsp.portal.covisint.com/documents/106025/14555722/FMEA+Handbook+v4.2/4c14da5c-0842-4e60-a88b-75c18e143cf7?version=1.0.

Table 3 Fmea scales and values (adapted from [12])

Severity (S)	1 Negligible impact on the level of services
	2 Slight decrease in the level of services
	3 Visible decrease in services
	4 Major deterioration in operating services
	5 Services are unavailable
Likelihood of occurrence (O)	1 Once in many years
	2 Once in many operating months
	3 Once in some operating weeks
	4 Weekly occurrence
	5 Daily occurrence
Likelihood of detection (D)	1 Very high/certain detection in a short time
	2 A number of warning in a short time
	3 Some warning
	4 Little warning
	5 Not detected

5. Evaluate the impact or gravity of the consequences (**S=Severity**).
6. the **Criticality** is determined by the combination of the likelihood of failure and the severity of the consequences.
7. Develop a strategy to bypass the failure mode, reduce the impact of failure, or take action if the failure happens. A characteristic is classified as critical when it is vital to conform to a certain requirement in order to prevent failure.

Those phases are also shown in Fig. 3. The qualitative scale for assigning the three components of risk value are often divided in 5, 7 or 10 different levels for both Severity, Probability of Occurrence and Detect. An example of a 5-level scale is shown in Table 3 a 10-level scale may be found in [8].

6 Privacy Impact Assessment Methodology by the French Data Protection Authority (CNIL-PIA)

The General Data Protection Regulation (GDPR), entered into force in 2018, at article 35 states that in several cases, in particular involving new technologies, a risk assessment procedure is mandatorily requested to data controllers (i.e. the management of company/public agency).

For this reason, the French Data Protection Authority, namely *Commission Nationale de l'Informatique et des Libertes* (CNIL), developed since 2018 the tool PIA, a risk assessment methodology particularly suited for privacy risk assessment [13]. The approach used by PIA methodology is qualitative, so relies on the subjective judgment of a team of experts responsible for determine the overall risk.

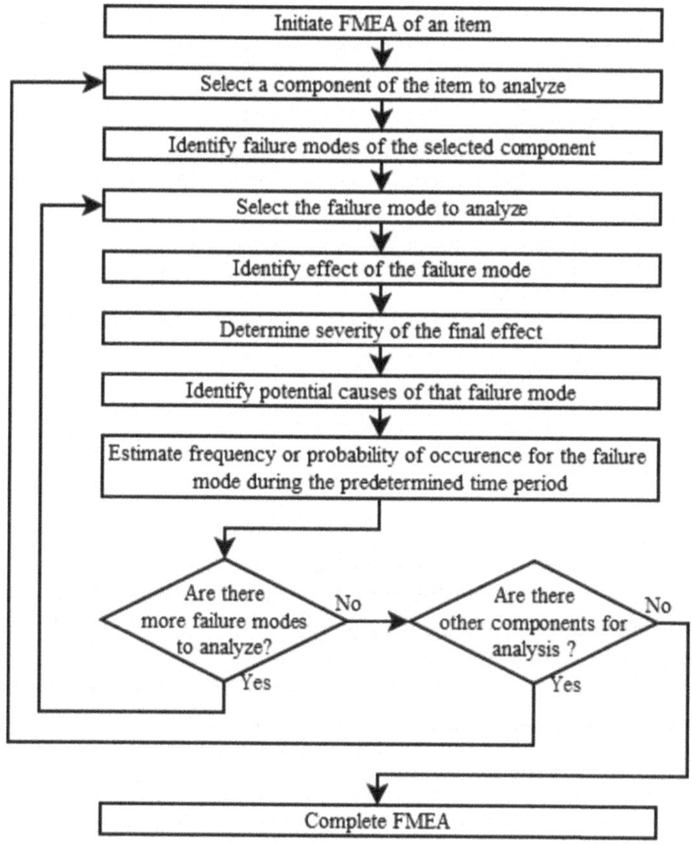

Fig. 3 Fmea flow chart (based on IEC 60812)

The PIA methodology necessitates assessing the potential consequences on individuals' privacy for each feared event (illegitimate access, unwanted change, and disappearance of personal data). This involves determining the impacts on data subjects' privacy if the event were to occur and estimating its severity, taking into account the harmful nature of the potential consequences and any controls that could mitigate them.

The scale utilized for measuring both severity and likelihood is founded on a set of four values, ranging from 1 (Negligible) to 4 (Maximum). A further description of this scale may be found in Table 4. In the methodology, three feared event are taken into account: (i) (I)—illegitimate access, (ii) (U) unwanted change and (iii) (D) disappearance of personal data, corresponding to the three elements of the CIA triad (Confidentiality, Integrity, and Availability). For each feared event, the methodology requires that:

Table 4 Risk metrics of PIA methodology

Level	Severity	Likelihood
1 Negligible	Data subject not be affected/few inconveniences	It seems (almost) impossible for the threat to materialize
2 Limited	Data subject affected by surmountable inconveniences	It is difficult for the threat to materialize
3 Important	Data subject affected by inconveniences that they can overcome though with true and serious difficulties	It is clearly possible for the threat to materialize
4 Maximum	Data subjects may encounter significant, or even irreversible, consequences, which they may not overcome	It seems extremely easy for the threat to materialize

- determine the potential **impacts** on data subjects' privacy if it occurred
- estimate its **severity**, particularly depending on the prejudicial nature of the potential impacts and, where applicable, controls that could change them;
- Identify the **threats** to personal data supporting assets that could lead to this feared event and the risk sources that could cause it;
- estimate its **likelihood**, particularly depending on the level of vulnerabilities of personal data supporting assets, the level of capabilities of the risk sources to exploit them, and the controls that could change them;
- Determine whether the risks identified in this way can be considered acceptable in view of the existing or planned controls (already undertaken) and, if not, propose additional controls and re-assess the level of each of the risks in view of the latter, so as to determine the residual risks.

The scale used for both severity and likelihood is based on four values, from 1 (Negligible) to 4 (Maximum) and it is described in the following Table 4.

The PIA tool uses also a graphic representation of the assessment procedures, similar to Risk Matrix. In the Fig. 4, the red dots are the values of assessment of the feared events (D, I and U). The background is colored in red (upper right side) and green (lower left side) to respectively represent high and low risk areas, while in the central area the green and red colors blend into each other. The blue points represent the values of the risk after a planned risk mitigation action.

For specific cases, you can find precise recommendations and illustrations about the rating for Severity and Likelihood in [15]. The papers [14, 16] provide a detailed explanation of a comprehensive risk assessment using PIA methodology in particular situations.

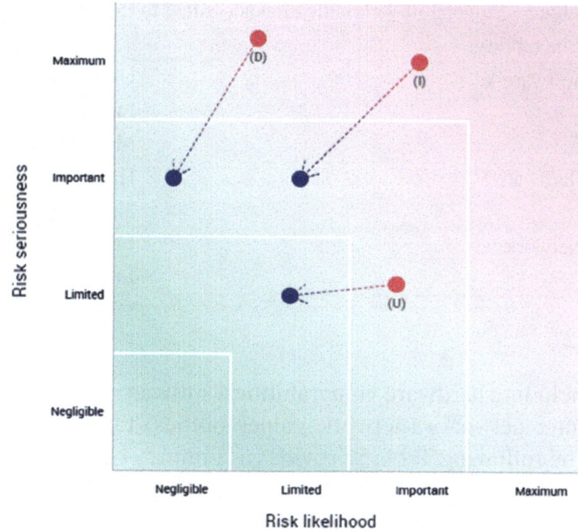

Fig. 4 The graphic representation of PIA assessment and risk mitigation (adapted from [14])

7 Risk Assessment in Cyber Physical Systems (CPS)

In order to incorporate ICT security into the analysis, it is imperative to define a cause-effect chain, so security-critical incidents can be categorized into comparable stages. The proposed components of a security cause-effect chain include the following factors:

- Vulnerabilities
- Threat Agent, Mode and Effect
- Attack Probability.

A complete survey on CPS security issues may be found in [1].

7.1 *Vulnerabilities*

A weak point or vulnerability is an indispensable prerequisite for a successful security breach of a system. A vulnerability is exactly a failure cause as defined in the previous Sections, and the identifying vulnerabilities is a fundamental requirement in risk analysis.

According to ISO/IEC 27002, a vulnerability is defined as a weakness in an asset or a group of assets that can be taken advantage of by a threat. ISO/IEC 27005 categorizes vulnerabilities related to information security into different categories,

Table 5 Vulnerability classification according to ISO/IEC 27005 (adapted from [18])

Vulnerability	
Software V.	User authentication
	Input handling
	Memory safety
Hardware V.	Hardware Trojan
	Non tamper-resistant HW
Network V.	Intercepted
	Interrupted

including hardware vulnerabilities, software vulnerabilities and Weaknesses in computer networks (network vulnerabilities) that can be exploited by attackers [17]. The following Table 5 provides a summary of a potential classification system using ISO/IEC 27005.

Weaknesses can be found at the network, hardware, and software levels. Hardware vulnerabilities pose a significant challenge for security engineering when components of embedded systems are used in an environment that may not be reliable or trustworthy. Furthermore, it is possible for hardware to be outfitted with supplementary harmful components. Re-configurable refers to the capability of a microcontroller to be reprogrammed. This may be due to the omission of a particular stage in the commissioning procedure. In the absence of tamper-proof hardware, a malicious individual could get unauthorized access to hardware components and carry out direct attacks on the hardware [19].

Other notable vulnerability classifications are the CWE (Common Weakness Enumeration), that is an exhaustive and collaboratively-created inventory of prevalent vulnerabilities in software. By employing a top-down methodology, vulnerabilities that exist at a lower level of design become more specialized. In Table 6 are listed the first ten of the *2023 CWE Top 25 Most Dangerous Software Weaknesses*.[2]

Examples of security analysis using CWE classification may be found in [17, 20, 21].

7.2 *Threats Agents, Mode and Effects*

A threat agent is an entity, whether it be an individual, group, organization, or government, that engages in or intends to engage in harmful acts. When a threat agent exploits a weakness, it jeopardizes the security of the system. Vulnerabilities that lack a threat agent do not result in any consequences and pose a minimal level of risk. A threat agent is an essential component of the cause-effect chain for ensuring safety.

[2] https://cwe.mitre.org/top25/archive/2023/2023_top25_list.html.

Table 6 The first ten Most Dangerous Software Weaknesses of the 2023 CWE list

Rank	ID	Name	Score
1	CWE-787	Out-of-bounds Write	63.72
2	CWE-79	Improper Neutralization of Input During Web Page Generation ('Cross-site Scripting')	45.54
3	CWE-89	Improper Neutralization of Special Elements used in an SQL Command ('SQL Injection')	34.27
4	CWE-416	Use After Free	16.71
5	CWE-78	Improper Neutralization of Special Elements used in an OS Command ('OS Command Injection')	15.65
6	CWE-20	Improper Input Validation	15.50
7	CWE-125	Out-of-bounds Read	14.60
8	CWE-22	Improper Limitation of a Pathname to a Restricted Directory ('Path Traversal')	14.11
9	CWE-352	Cross-Site Request Forgery (CSRF)	11.73
10	CWE-434	Unrestricted Upload of File with Dangerous Type	10.41

Threat mode categorizes the manner in which vulnerabilities are leveraged. Exploitation of vulnerabilities can occur through diverse methods, each yielding distinct consequences and requiring certain conditions. The potential modes of threat rely on the characteristics of the system and the abilities of the threat agent. Threat modes can range from basic actions like disrupting a connection to more complex operations such as taking advantage of a vulnerability in the input system by delivering a well crafted input signal. Typically, this can be correlated with a breach of a security attribute. The correlation between security attributes and system quality attributes differs for each specific system. The impact of each threat mode on the dependability attributes (Reliability, Availability, Maintainability, and Safety) may vary depending on the system. One widely used framework for classifying threats is STRIDE [22]. As indicated in Table 2, an exploited vulnerability results in one of the following generic consequences.

A threat mode is analogous to the failure mode as explained in the previous Sections, and delineates the specific manner in which the security measures become ineffective. To classify threat and failure modes, one could utilize the approach outlined in Ref. [19]. Various characteristics of failure modes are defined and categorized.

The threat effect refers to the outcome or result in terms of the operation, function, or status, and is analogous to the failure impact in the field of safety. The threat mode refers to the specific security property that has been compromised, while the threat effect refers to the system quality attribute that has been violated. Violated qualities extend beyond security. All reliability characteristics may be impacted. The specific attribute that is violated in a particular circumstance relies on the system, its surroundings, and its operating condition.

7.3 Attack Probability

The attack probability refers to the likelihood of hardware, software or network failure. In other words, it is the probability of a threat agent successfully carrying out the intended threat impact in a security context. The level of dependence is contingent upon both the threat agent's characteristics and the features of the system, as well as the system's environment. The likelihood of a successful assault is quite low if a system is not connected to a public network and is situated in a confined area. Furthermore, apart from the technical likelihood of an assault, each threat actor possesses distinct motivational motivations and capabilities. Capabilities refer to the combination of financial resources, knowledge, and maybe other resources possessed by a threat agent, which are utilized to exploit vulnerabilities. The threat agent is characterized by their motivation and capability, and the sum of these factors determines the "Threat property" parameter, as shown in Formula 3.

$$ThreatProperties = Motivation + Capability \tag{2}$$

In this formula the Motivation scale is: (1) = opportunity target, (2) = mildly interested, (3) = main target, and capability (of the Threat agent) scale is: (1) = low, (2) = medium, (3) = high.

Aside from the characteristics of the threat agent, several system variables also impact the likelihood of an assault. Reachability is quantified on a scale of 1 to 3 and represents the level of ease with which one can establish a connection to the system. Examples of systems that fall within reachability 3 are those that are directly connected to the internet and may be easily discovered using common tools. If a system is not directly connected to the internet but can be accessed through a network that is connected to the internet, it is granted a reachability score of 2. Systems that are completely disconnected from any network have a reachability score of 1. Aside from reachability, another determinant of a system's vulnerability is the atypicality of its components and designs. It is reasonable to infer that prospective threat actors possess limited understanding about uncommon systems, resulting in a greater amount of work required to identify vulnerabilities and exploit them. The combined properties determine the parameter "System's susceptibility".

$$SystemSusceptibility = Reachability + Unusualness \tag{3}$$

In the formula the scale for Reachability is: (1) = no network, (2) = private network, (3) = public network, and for Unusualness is: (1) = restricted, (2) = commercially available, (3) = standard.

The sum of Threat properties and System susceptibility gives a measure of the risk probability. Values of attack probability from 1 to 4 are to be considered low, from 8 to 12 high and in the gray area risk effects have to be well estimated.

8 Conclusion

In conclusion, it is possible to argue that in this field additional research is necessary to provide a dependable evaluation of the risk associated with security concerns. It is required specifically verified facts regarding the likelihood or frequency of attacks. This would enable the calibration of the severity of security threats in order to achieve results that can be compared with the severity of safety-related issues. The current technique, which utilizes a semi-quantitative method to differentiate between four parameters that contribute to the probability of an attack, is an initial step in addressing this difficulty. These factors are categorized into system properties and attacker properties, both of which have an impact on the likelihood of an attack occurring. Further study is needed to include inmate threat agents in the analysis, as the current attacker properties only apply to human threat agents.

References

1. Alguliyev R, Imamverdiyev Y, Sukhostat L (2018) Cyber-physical systems and their security issues. Comput Ind 100:212–223
2. Campanile L, Iacono M, Mastroianni M (2022) Towards privacy-aware software design in small and medium enterprises. In 2022 IEEE international conference on dependable, autonomic and secure computing, International conference on pervasive intelligence and computing, International conference on cloud and big data computing, International conference on cyber science and technology congress (DASC/PiCom/CBDCom/CyberSciTech), pp 1–8. https://doi.org/10.1109/DASC/PiCom/CBDCom/Cy55231.2022.9927958
3. Harkat H, Camarinha-Matos LM, Goes J, Ahmed HF (2024) Cyber-physical systems security: a systematic review. Comput Ind Eng 188:109891
4. Brunner M, Sauerwein C, Felderer M, Breu R (2020) Risk management practices in information security: exploring the status quo in the dach region. Comput Sec 92:101776
5. Shameli-Sendi A (2020) An efficient security data-driven approach for implementing risk assessment. J Inf Sec Appl 54:102593. https://doi.org/10.1016/j.jisa.2020.102593
6. Shameli-Sendi A, Aghababaei-Barzegar R, Cheriet M (2016) Taxonomy of information security risk assessment (isra). Comput Sec 57:14–30
7. Kovacevic N, Stojiljkovic A, Kovac M (2019) Application of the matrix approach in risk assessment. Oper Res Eng Sci Theory Appl 2(3):55–64
8. Damanab PS, Alizadeh SS, Rasoulzadeh Y, Moshashaie P, Varmazyar S (2015) Failure modes and effects analysis (fmea) technique: a literature review. Sci J Rev 4(1):1–6
9. Cândea G, Kifor S, Constantinescu C (201) Usage of case-based reasoning in fmea-driven software, Procedia CIRP 25: 93–99. 8th International conference on digital enterprise technology - DET 2014 Disruptive Innovation in Manufacturing Engineering towards the 4th Industrial Revolution. https://doi.org/10.1016/j.procir.2014.10.016. https://www.sciencedirect.com/science/article/pii/S2212827114010476
10. Sharma KD, Srivastava S (2018) Failure mode and effect analysis (fmea) implementation: a literature review. J Adv Res Aeronaut Space Sci 5(1–2):1–17
11. Spreafico C, Russo D, Rizzi C (2017) A state-of-the-art review of fmea/fmeca including patents. Comput Sci Rev 25:19–28
12. Christopher M (2022) Logistics and supply chain management. Pearson UK
13. French Data Protection Authority (CNIL), Privacy Impact Assessment (PIA) - Methodology (2018). https://www.cnil.fr/sites/default/files/atoms/files/cnil-pia-1-en-methodology.pdf

14. Campanile L, Iacono M, Marulli F, Mastroianni M (2021) Designing a GDPR compliant blockchain-based IOV distributed information tracking system. Inf Process Manag 58(3):102511
15. French Data Protection Authority (CNIL), Privacy Impact Assessment (PIA) - Knowledge Bases (2018). https://www.cnil.fr/sites/default/files/atoms/files/cnil-pia-3-en-knowledgebases.pdf
16. Mastroianni M, Palmieri F, Ficco M, Kozik R, ChoraÅ M (2023) Privacy risk analysis and metrics in capturing and storing network traffic. In: 2023 24th international conference on control systems and computer science (CSCS), pp 580–585. https://doi.org/10.1109/CSCS59211.2023.00097
17. Wang T, Qin S, Chow KP (2021) Towards vulnerability types classification using pure self-attention: A common weakness enumeration based approach. In: 2021 IEEE 24th international conference on computational science and engineering (CSE), pp 146–153. https://doi.org/10.1109/CSE53436.2021.00030
18. Schmittner C, Gruber T, Puschner P, Schoitsch E (2014) Security application of failure mode and effect analysis (fmea). In: Bondavalli A, Di Giandomenico F (eds) Computer safety, reliability, and security. Springer International Publishing, Cham, pp 310–325
19. Lyu X, Ding Y, Yang S-H (2019) Safety and security risk assessment in cyber-physical systems. IET Cyber-Phys Syst: Theory Appl 4(3):221–232. https://doi.org/10.1049/iet-cps.2018.5068
20. Galhardo C, Mell P, Bojanova I, Gueye A (2020) Measurements of the most significant software security weaknesses. In: Proceedings of the 36th annual computer security applications conference, ACSAC '20, Association for Computing Machinery, New York, NY, USA, pp 154–164. https://doi.org/10.1145/3427228.3427257
21. Malik AA, Tosh DK (2021) Robust cyber-threat and vulnerability information analyzer for dynamic risk assessment, in. IEEE Int Med Conf Commun Netw (MeditCom) 2021:168–173. https://doi.org/10.1109/MeditCom49071.2021.9647584
22. Xiong W, Lagerström R (2019) Threat modeling - a systematic literature review. Comput Sec 84:53–69

Healthcare and Cyberspace: from Cyber-Physical Systems to Medical Digital Twins

Rocco Zaccagnino, Delfina Malandrino, Gianluca Zaccagnino, Gerardo Benevento, and Margherita Maria Napolitano

Abstract Cyber-Physical Systems (CPS) are crucial for developing Digital Twins (DT) due to their ability to bridge physical systems and virtual components. CPS integrate computational and physical processes with continuous feedback, traditionally analyzed through costly facilities. High-fidelity virtual models now serve as the basis for DT, which is emerging as a pivotal technology to realize CPS functions, such as to create bidirectional interactions between the physical and cyber realms. DT is driving significant progress on healthcare, where the abundance of data collected from numerous IoT devices and AI models enables: *(i)* the monitoring of patients' lifestyles and regular eating habits and, *(ii)* providing alerts to patients about prescriptions and medical consultations, and *(iii)* assisting in selecting appropriate medications and predict the outcomes of surgical procedures. The healthcare sector increasingly demands advanced techniques to offer patients optimal assistance and drive digital transformation initiatives forward. This study seeks to elucidate and organize these guidelines, specifically aiming to clarify the integration of DT in healthcare. Finally, starting from an existing solution in literature, we propose an innovative preliminary medical model that relies on *Geometric Deep Learning* for the patient representation (the Patient DT) and *Ontologies* to facilitate the translation of complex healthcare data into a graph-based structure used by the Patient DT.

1 Digital Twin Towards Healthcare

A DT [27] is a virtual replica of a product, system, or person that, utilizing real-time data, aims to enhance decision-making through learning, reasoning, and dynamic recalibration mechanisms. At the core of DT is the coexistence of the physical space, representing the real environment, and the *virtual space*, managing all information

R. Zaccagnino (✉) · D. Malandrino · G. Benevento · M. M. Napolitano
Department of Computer Science, University of Salerno, Fisciano, SA, Italy
e-mail: rzaccagnino@unisa.it

G. Zaccagnino
TopNetwork SPA, Via Simone Martini 143, 00142 Roma, Italy

M. Ficco and G. D'Angelo (eds.), *Artificial Intelligence Techniques for Analysing Sensitive Data in Medical Cyber-Physical Systems*, Engineering Cyber-Physical Systems and Critical Infrastructures 12, https://doi.org/10.1007/978-3-031-70775-9_4

related to the *physical space*. The DT establishes a connection between the information in the two spaces throughout the life cycle of the simulated entity. As of now, the primary sectors effectively leveraging DT include those associated with the manufacturing sector, such as Industry 4.0, with the healthcare sector standing out as one of the fields benefitting the most from this technology.

Currently, the abundance of medical data and available technologies appears to make the creation of a digital model of the human body more feasible, enabling the testing of treatment effects or allowing surgeons to practice before an operation. However, the development of the *"digital human body"* faces a range of technical, ethical, and privacy issues that effectively hinder its progress, such as the precise modeling of human organs. Achieving this goal through DT technology would empower doctors to enhance the quality of life for patients by providing precise guidance on disease prevention [28–30]. Simultaneously, patients can use simulations to proactively identify signs of disease and recognize potential disease recurrence. DTs used in the healthcare sector extensively employ AI, ML, and Cloud technologies, aiming to utilize real-time data for creating accurate health predictions [31–33]. The most intriguing aspect of the healthcare DT paradigm is its *interactivity*, where patients actively participate in the DT, promoting transparency and trust throughout the therapy process.

Healthcare faces various challenges in adopting DT technologies [34, 35]. The global healthcare sector has demonstrated resilience in overcoming obstacles posed by COVID-19. However, the profound effects of the pandemic have presented significant challenges to hospitals and patients due to unexpected workload changes [36–38], enabling millions of people worldwide to swiftly access life-saving treatments. Another notable aspect that has yielded promising results from DT technologies is their *collaborative* use. Constructing a high-fidelity DT necessitates the collaboration of various parties, including hospital managers, doctors, administrative staff, and technological partners [39].

1.1 Features and Services of DT for Healthcare

Advancements in human body modeling are progressively enhancing the efficiency of clinical operations and hospital administration through DTs, making them faster and less expensive. This improvement contributes to enhanced medical care through digital monitoring [40–42]. Figure 1 illustrates smart services offered by DT in the healthcare sector, with a focus on services centered around patient health [43–45].

In summary, the primary objective of every healthcare organization is to enhance the patient experience, ensuring high effectiveness, productivity, and quality of care. Precise data management necessitates the capability to integrate data from diverse sources, including electronic health records, disease registries, and notably, patient data from wearable and other healthcare devices. This data can be employed to

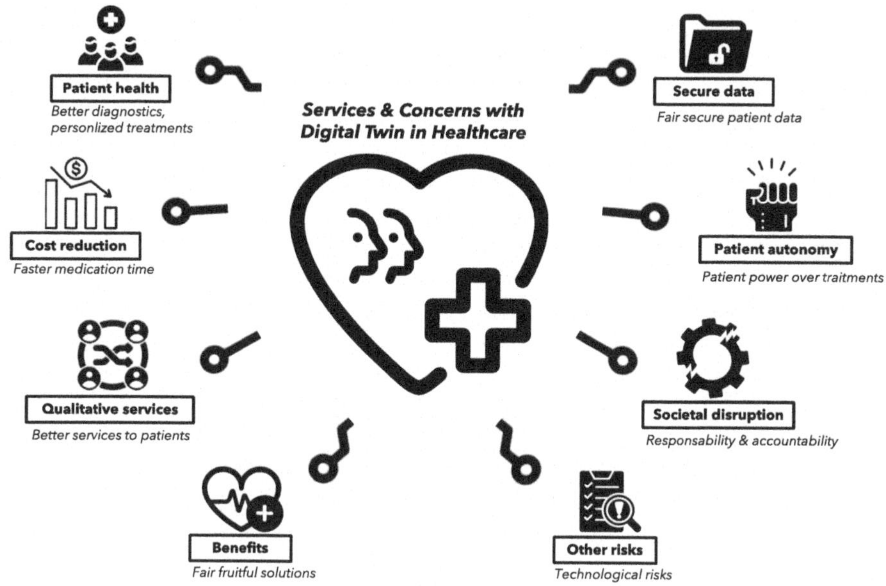

Fig. 1 DT in healthcare: services & concerns

conduct simulations, predict the effects of different drugs, formulate personalized medical treatment plans, and provide early warnings about patient susceptibility to diseases/disorders.

To ensure these services, DT mainly uses four technologies to create visual representations, collect, store, and analyze data, and provide insights: AI, IoT, Cloud, and Augmented Reality. Using such technologies, a DT constantly *(i)* monitors the behavior of physical equipment, *(ii)* runs an asset behavior prediction algorithm to increase efficiency and profitability, and *(iii)* performs risk analysis on hypothetical situations that allow the medical organization to maximize operational efficiency.

1.2 Digital Twin Applications for Healthcare

As of now, numerous healthcare scenarios incorporate applications in which DTs play a pivotal role. Below, we will delve into the most significant applications of DT in healthcare:

- Patient care. The pinnacle goal of DT is likely represented by personalized medicine. Through DT, virtual simulations of human organs enable preliminary testing before the actual implementation of treatments on patients. Additionally, DTs, by providing virtual representations of individual patients, facilitate predictive maintenance of resources such as laboratory automation and surgical robots, thereby enhancing their performance, accessibility, and cost-effectiveness.

- *Clinical studies.* DTs have the potential to revolutionize the management of clinical trials. By creating a DT for each participant, the number of required participants can be reduced through virtualization of the test group. Integration of IoT devices and telemedicine procedures optimizes patient participation, allowing virtual engagement and expanding the geographical diversity of participants.
- *Enhancement of hospital activities.* DTs can simulate various hospital operations and entities, including staff organization, patient management, and equipment usage. Utilizing an AI-based model with real-time data from all hospital data points and laboratory diagnosis results, predictions and analyses are provided to medical staff, facilitating decision-making. This involves defining disease models, treatment effects, and choices for patients with similar characteristics.
- *Diagnostics and medical training.* DTs can simulate patients before actual surgeries, enabling virtual surgeries by multidisciplinary teams to prevent collateral damage. As a study tool, DTs allow trainees to mimic surgical procedures on virtual patients. The integration of advanced technologies like AI into DTs significantly benefits medical education, fostering widespread use of technology to enhance healthcare.
- *Healthcare forecasting.* The notable increase in the demand for DT in the medical sector is driven by the promise of tailored treatments based on health predictions. Researchers can run numerous personalized treatment simulations using patient information such as genetic makeup and medical history. This not only helps identify the most effective therapy approach without endangering actual patients but also facilitates the optimization of hospital activities and resource distribution.
- *Remote patient assistance.* Through smart wearable devices, doctors can remotely monitor patients. Real-time data from cloud-based DTs is transmitted through wearable devices equipped with sensors, enabling the coordination of equipment and personnel. The data derived from IoT and DT applications in healthcare plays a crucial role in the growth of healthcare and medical companies, aiding organizations and experts in determining optimal methods to enhance processes, improve patient experience, reduce operating costs, and increase treatment value.

2 Digital-Twins and Cyber-Physical Systems: Opportunities for Integration

Innovative technologies in industrial production, such as IoT, Cloud Computing, Big Data Analytics, and AI, are ushering in a transition from knowledge-based intelligent manufacturing to data-driven intelligent manufacturing. In this context, "intelligent" refers to how data are generated and utilized. This shift involves intelligent production based on real-time data transmission and analysis, coupled with simulation based on digital models for optimization throughout the entire product life cycle. The foundational requirement for realizing this vision of intelligent production is the *digital-physical integration*.

As discussed above, within the progression of industrial technologies, two types of technologies facilitating the integration of digital and physical systems have gained prominence: CPS and DT. Both of these technologies have sparked significant interest among researchers and professionals. To better understand the benefits derived from their integration, let's first revisit their definitions. A DT is a collection of virtual information constructs that comprehensively describe a potential or existing physical product. On the other hand, CPS represents a new generation of systems with integrated computational and physical capabilities, capable of interacting with humans through various modalities. The central challenge for achieving comprehensive digitization lies thus in "cyber-physical integration", and in this context, both CPS and DT emerge as potent tools to realize this objective.

2.1 Inside CPS and DT

Differences and Similarities

While, by definition, both CPS and DT involve the coexistence of two major spaces, the digital and the physical, the distinctions lie in the objectives for which these spaces are managed. CPS, defined as the integration of physical and computational processes, aims to enhance physical systems' functionality through the application of computation, communication, and control (3Cs). On the other hand, DT utilizes the digital space to simulate the physical system, with the goal of optimizing its life cycle in real-time. A key point of comparison between these technological models is the emphasis on specific elements in implementing their functions. CPS places sensors and actuators at the forefront, focusing on the "exchange of information" with the physical space, enabling the perception of physical systems and control activities. In contrast, DT prioritizes the digital model and the data it incorporates. Given the dynamic nature of the physical space, the digital model, supported by sensors and devices, strives to conduct more "accurate analyses and predictions" to enhance decision-making processes.

Integration

The integration of computational and physical processes in CPS involves monitoring physical processes through integrated computers and networks, employing feedback mechanisms where physical processes influence computations and vice versa. Shifting the focus to the optimal management of the physical model and generated data allows for the replacement of expensive physical plants with high-fidelity virtual models on which DT can be built.

If CPS aims to establish two-way interaction channels between the physical and cyber worlds, creating an Internet of Things (IoT) in the physical realm by connecting various sensors, actuators, and controllers with perception products and equipment

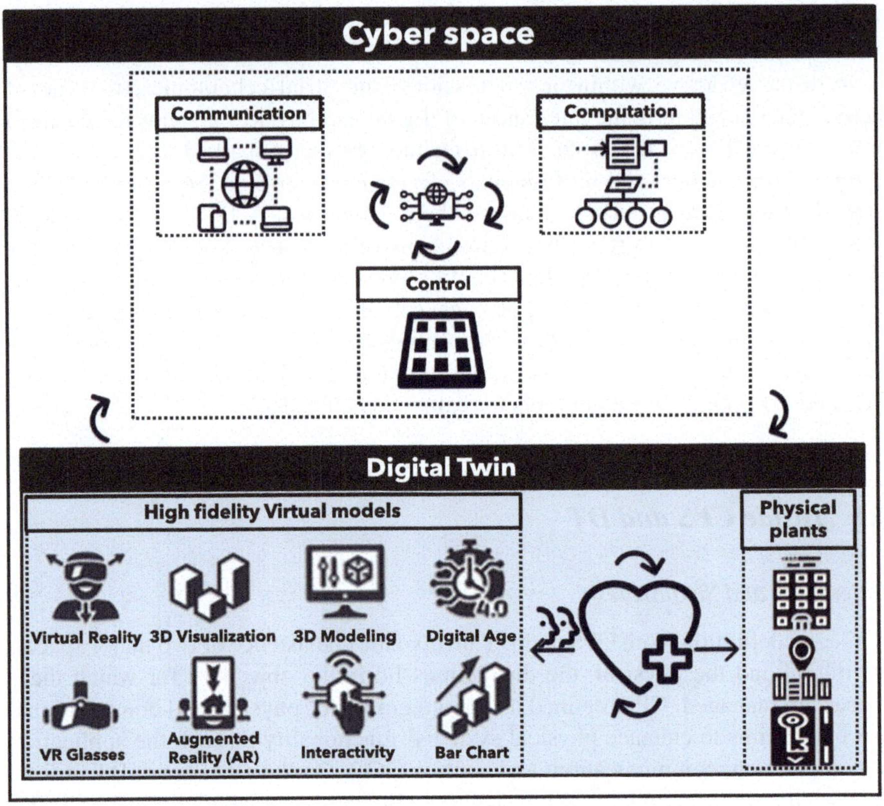

Fig. 2 CPS and DT integration

for real-time data transmission, processing, feedback, and services, then DT emerges as a crucial simulation-based technology to fulfill these functions. The connection between CPS and DT is illustrated in Fig. 2. Through the application of 3Cs technologies (computation, communication, and control) [46, 47], CPS provides accurate control and independent management for physical functions. Simultaneously, the encapsulated DT establishes an interconnection between physical plants and high-fidelity virtual models.

The development and implementation of a realistic DT for CPS often necessitate the application of methodologies grounded in Computational Intelligence (CI). Simulations of real-world phenomena tend to be computationally expensive due to reliance on intricate mathematical models that pose challenges for efficient optimization techniques [48–50]. To address this, researchers commonly employ CI methods, such as evolutionary computing, fuzzy logic, and artificial neural networks, to enhance the efficiency and effectiveness of the simulation process.

2.2 CPS and DT for Healthcare: Applications, Safety and Security

The adoption of Medical Cyber-Physical Systems (MCPS) in healthcare settings is crucial for ensuring high-quality patient care, but it also brings forth various challenges. Security, privacy, and reliability are paramount considerations in the development and implementation of MCPS. Researchers have explored innovative applications of CPS in healthcare, such as an artificial pancreas for blood glucose level control and a post-stroke monitoring system utilizing wearable sensors and ML algorithms [51, 52].

The concept of digital twins has been introduced to enhance security measures for CPS. DTs are seen as a potential holistic approach to CPS security, offering a way to implement security-enhancing technical solutions [53–56]. Cyber-Physical-Security emerges as a term that highlights the integration of CPS with security systems. In-depth investigations have been conducted to identify and address security threats in CPS, both in public and private sectors [57]. Safety, security, and sustainability are identified as essential qualities that CPS must guarantee, encompassing the ability to avoid dangers, ensure the authenticity, integrity, and secrecy of information, and the ability to operate sustainably with renewable energy sources.

In the healthcare context, the security of MCPS is a critical aspect, particularly in pervasive healthcare monitoring systems. Methodologies for analyzing the security characteristics of closed-loop medical device systems have been proposed to address the unique challenges in ensuring the security of healthcare-related CPS [58, 59].

2.3 CPS and DT for Healthcare: a Systematic Mapping

In this section, we provide the mapping of recent literature concerning CPS and DT for Healthcare [76–79]. To perform such a mapping, we have used the following search query on the Scopus[1] database:

```
TITLE-ABS-KEY ("digital twin" AND ("health" OR "healthcare"
OR "medical" OR "medicine") AND ("cyber-physical" OR "cyber
physical")) AND PUBYEAR > 2020
```

We have also restricted the article set to those written in the English language. The crawled papers' dataset has been exported in the `bibtex` format and analyzed using *Bibliometrix* [80], an R open source package, for performing comprehensive science mapping analysis. This package provides a full set of analytical tools suited for analyzing bibliographical data originating from several abstract databases. From the analysis, we extracted several insights here below reported.

In Table 1 we show the key information about the dataset of identified papers. In particular, in the time span between 2021 and 2023, 68 documents have been

[1] http://scopus.com.

Table 1 Key information about the dataset of identified papers

Key information	Value
Timespan	2021 : 2023
Sources (Journals, Books, etc.)	58
Documents	68
Annual growth rate %	–2.99
Document average age	1.01
Average citations per doc	5.75
Average citations per year per document	2.755
References	3134

selected coming from 58 different sources (journals, books, conferences, etc.). When analyzing the category sources breakdown, we found an interesting insight, that 24% of them are review papers. The remaining are 44% articles, 6% book chapters, and 26% conference papers.

Most of the production comes from China and UK as shown in Fig. 3.

The annual scientific production has grown from 2021 to 2022 (from 17 to 35 published papers), but it seems (although some upcoming papers should be available by the end of 2023) braking in 2023, with just 16 published papers. The relevant sources are shown in Fig. 4: we can note that five of the identified papers come from the Communications in Computer and Information Science book series by Springer.

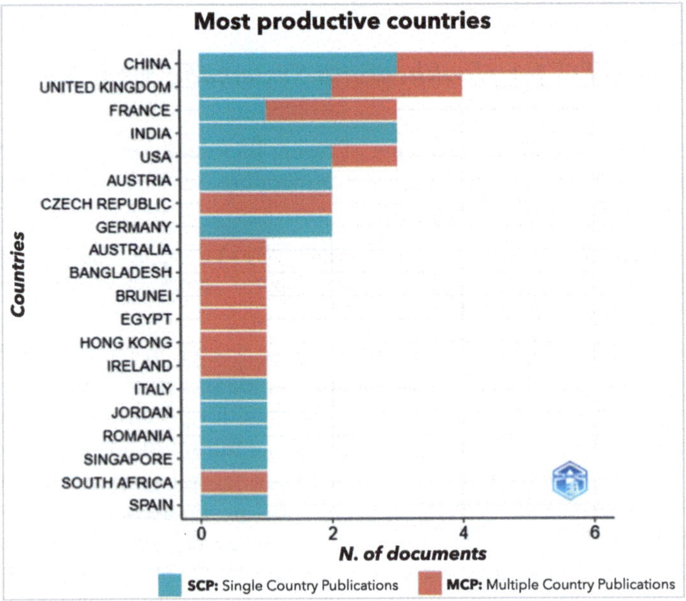

Fig. 3 Most productive countries in terms of annual scientific production

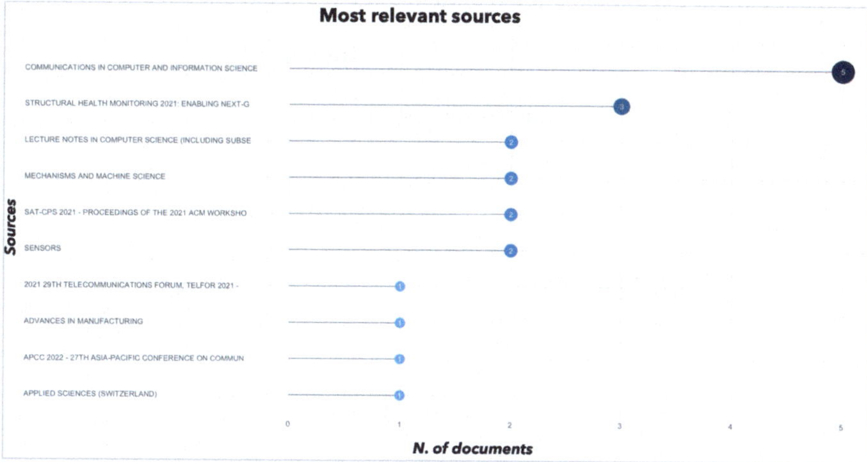

Fig. 4 Most relevant sources

Figure 5 shows the thematic map based on co-word network analysis. The words used are "Indexed Keywords". These are keywords chosen by content suppliers and are standardized based on publicly available vocabularies. Unlike Author keywords, the Indexed keywords account for synonyms, various spellings, and plurals.[2] For this reason, Bibliometrix uses Indexed keywords for Scopus entries. To generate this chart, Bibliometrix uses a methodology borrowed from the methodological framework so-called thematic analysis [82]. The library computes the co-occurrence matrix of keywords in the paper and applies the Louvain algorithm to cluster the words of the graph. For each detected cluster, the library measures the Callon centrality and Callon density [83]. According to [84], higher values of centrality and density define the *Motor Themes*, well-developed and relevant for structuring the conceptual framework of the domain; higher values of centrality and lower values of density define the *Basic Themes*, significant for the domain and cross-cutting to its different areas; lower values of centrality and density define *Emerging or Declining Themes*, not fully developed or marginally interesting for the domain; lower values of centrality and higher values of density define *Niche Themes*, strongly developed but still marginal for the domain under investigation. Four different areas can be located through this figure, i.e., *Niche Themes*, *Motor Themes*, *Basic Themes*, and *Emerging or Declining Themes*.

Among the *Motor Themes* we find deep learning, followed by embedded systems at the boundaries between *Basic* and *Motor* themes; instead, fault detection, now placed in *Emerging or Declining Themes*, is going to be a *Basic theme* in the next future. We can compare these results with the ones in Figs. 6 and 7. It can be derived that data-driven solutions powered by AI and ML are going to be soon an *Emerging theme* in this area.

[2] https://service.elsevier.com/app/answers/detail/a_id/21730/supporthub/scopus/.

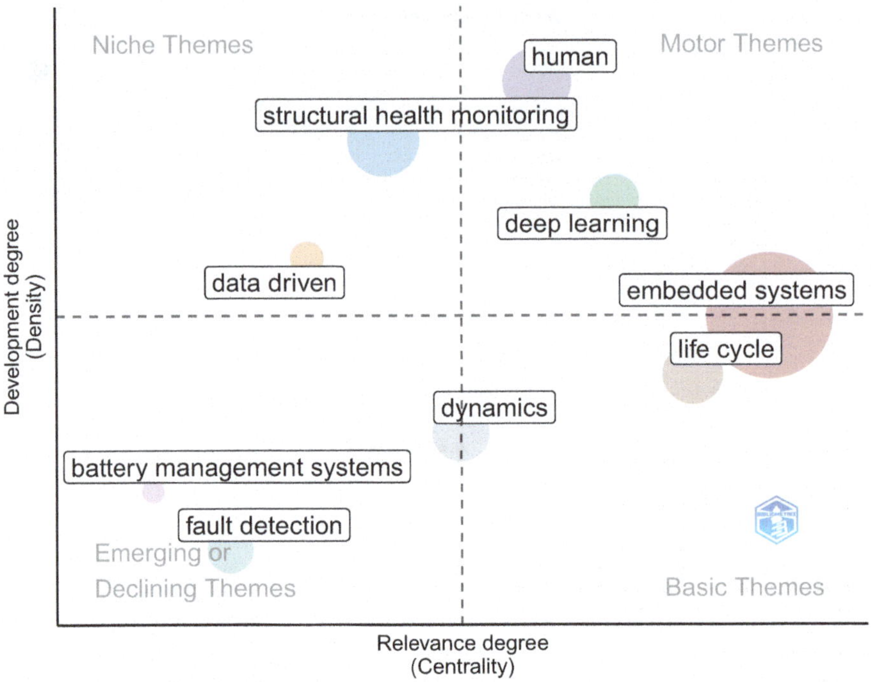

Fig. 5 Thematic map based on Indexed Keywords

Fig. 6 Treemap of Indexed Keywords

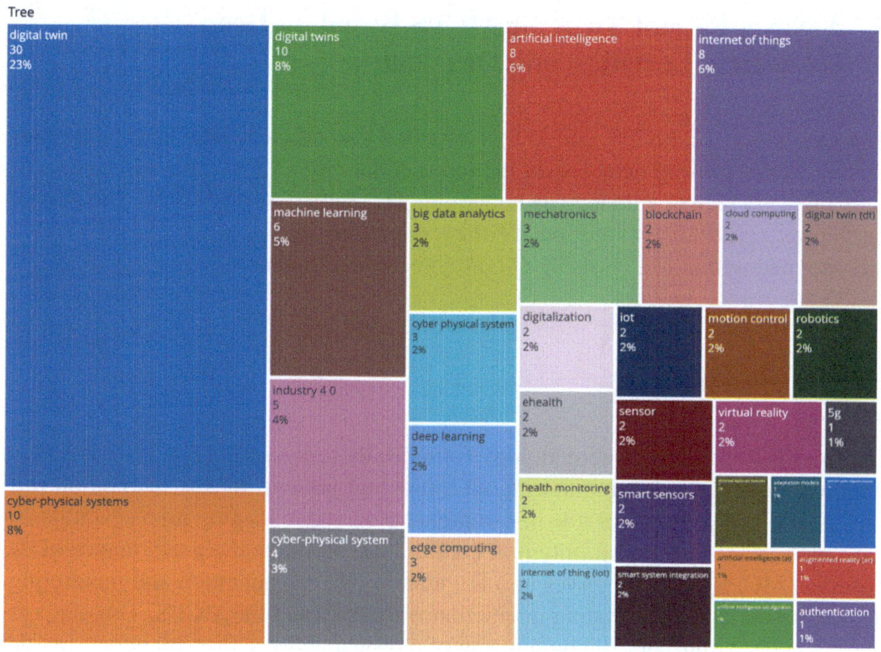

Fig. 7 Treemap of Authors' Keywords

Another interesting finding in this systematic mapping is that the keyword *trust* is often available in research papers proposing DT solutions together with *human* and *accident prevention*, proving that research in CPS and DT for Healthcare is putting the human at the center of the development (see Fig. 8).

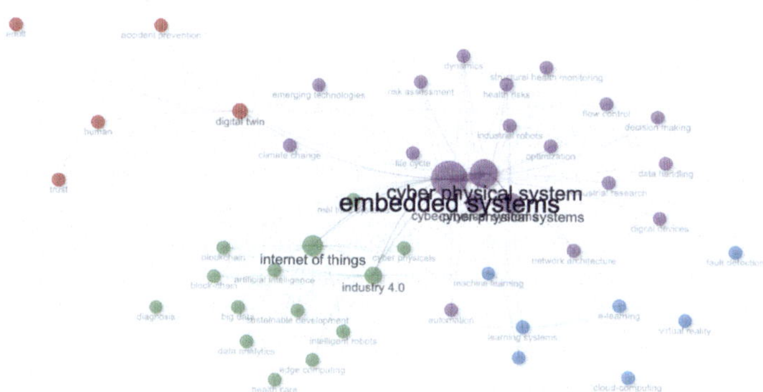

Fig. 8 Co-occurrence keywords' network

3 A Novel Patient Twin for MCPS Based on Deep Geometric Learning and Ontologies

The integration of MCPS and DT is essential for efficient healthcare solutions. Despite the increasing need for digital representations of CPS, especially in healthcare, there is a lack of guidance on integrating DTs with observed CPSs. This section introduces a preliminary solution: a *Patient DT* based on *Geometric Deep Learning*, incorporated within MCPS using *Ontologies* for health data.

3.1 Introduction

Our goal is to develop a methodology for digitally representing MCPS using DTs, with a specific focus on monitoring an individual *patient*. In this integration process, the DT plays a crucial role in comprehending and managing representations of data and services, aiming to monitor, control, and optimize the performance of the observed MCPS. The DT serves as a foundation for representation and monitoring, displaying data, and facilitating interaction with users and the MCPS. This supportive mechanism enhances the monitoring of physical entities, thereby improving the analysis of physical conditions, the application of personalized treatments, and the prediction of disease progression.

Two key aspects are considered in the design of our Patient DT for MCPS: *(i)* the *computational model* and *(ii)* the *integration model*.

The *computational model* plays a pivotal role in simulating various facets of the patient, addressing challenges presented by health data, such as high dimensionality, sparsity, limited quality, and fragmentation due to factors like acquisition technologies and security/privacy protections. Successful health representational models have been achieved by integrating information about health *connected* vital signs into Deep Learning (DL) methods [61]. Among these approaches, *Geometric DL* has emerged, concentrating on modeling and learning data connections and interactions as entities and relationships within a graph. Geometric DL employs specialized models called *Graph Neural Networks* (GNNs). Despite the impressive performance of GNNs in diverse fields, there have been limited endeavors to use them for defining a patient twin, and those that exist are still in a preliminary state, not fully integrated into a functioning Cloud infrastructure. This can be attributed to the non-trivial nature of graph-based modeling required for health data. However, a notable advantage of GNNs lies in their interpretability compared to other neural approaches, thanks to their graph structure [62].

The *integration model* is pivotal for integrating the Patient DT into the MCPS. The configuration of the MCPS can vary depending on the healthcare context, managed devices, medical personnel, and health information (vital signs). It is crucial to adapt the definition of a Patient DT to accommodate these variations. In our proposed preliminary solution, ontologies will be employed to handle healthcare data types and their relationships. The system will dynamically utilize the ontology to populate the GNN that simulates the human body. This approach shifts the designer's focus to modifying and managing the ontology rather than manually populating a powerful yet complex ML model like the GNN. We remark that the proposed Patient DT is based on the one proposed in [61].

3.2 System Architecture

In this section, we provide details of the MCPS framework within which the proposed Patient DT will be placed. The idea is to support such a Patient DT with a healthcare monitoring GNN model which automatically *(i) monitors* the clinical conditions of a patient, and *(ii) detects* abnormal physical health values by carefully monitoring clinical conditions through sensors deployed in an intelligent medical environment. The system architecture of our proposed MCPS framework is shown in Fig. 9.

Our framework is made up of three modules connected to each other:

- *Data module*: first the real-time data from the physical sensors are read and then pre-processed in order to further extract the features required for our system.
- *Patient DT ontology*: we defined healthcare ontology to provide information on entities and relationships (e.g., health data, vital signs) managed within the intelligent patient healthcare ecosystem.
- *Patient DT module*: the patient DT contains two sub-modules, i.e., the *GNN model* and the *health monitoring module*; the GNN is dynamically populated using the

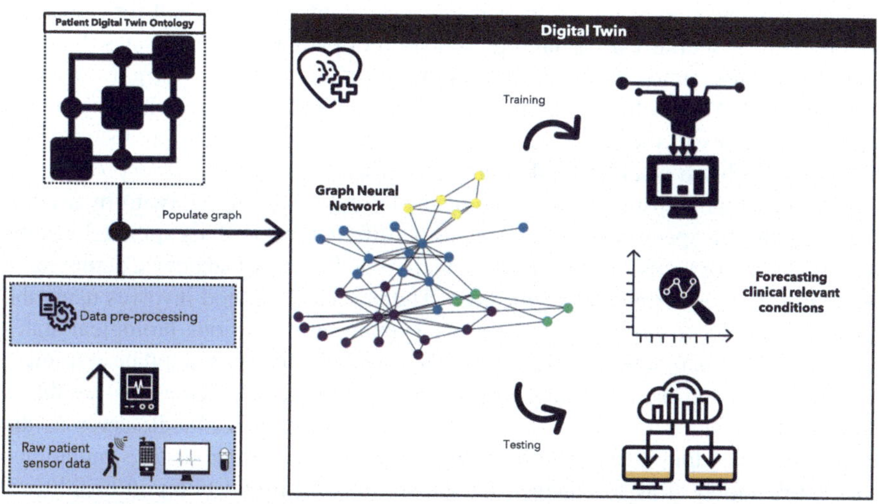

Fig. 9 Architecture of the proposed MCPS. A health ontology is used to dynamically populate the GNN model to forecast clinically relevant endpoints

data and related entities provided by the ontology module in real time; if necessary, i.e. if the available information changes over time, the GNN is retrained, and finally used by the health monitoring module to monitor the patient's health conditions.

3.3 The Patient DT

Overview

The changing landscape of healthcare underscores a transition from a reactive to a proactive and personalized model. Precision medicine, or *personalized medicine*, seeks to customize medical care based on individual characteristics, encompassing elements like genetics, lifestyle, and environmental influences. This paradigm shift is essential for delivering more precise and targeted treatments, ultimately enhancing patient outcomes. The integration of diverse data sources, ranging from genomic information to clinical history, plays a critical role in this endeavor. The challenge lies in thoroughly understanding and adapting to the intricate complexities of the human body [63, 64]. As technological advancements empower the collection and analysis of diverse datasets, the potential to revolutionize healthcare becomes increasingly apparent. Precision medicine's focus on individualized care, early disease detection, and targeted interventions aligns with the broader objective of improving healthcare efficiency. By considering multiple factors at various scales, including genomics and clinical information, healthcare providers can deliver more accurate diagnoses and personalized treatment plans. This approach holds the promise of enhancing both the quality and efficiency of healthcare delivery.

The core objective of this study is to introduce a proof of concept for a "patient digital twin," essentially a virtual representation of patients that simulates the intricate biological systems by integrating information at the levels of organs, tissues, and cells [65]. The profound complexity of the human body poses challenges, and existing solutions in the literature often focus on specific aspects. In a study conducted by Barbiero et al. [61], the authors advocate for generative approaches to graph representation [66, 67] to overcome the limitations of digital twins by accommodating the full spectrum of body signals at different levels. This approach results in a novel class of GNN-based ML tools, well-suited for medical device deployment and real-time monitoring and verification. The proposed method involves using the generative model to produce synthetic data representing various biological states that may not be easily observed in reality. This synthetic data can simulate extremely rare clinical scenarios, providing insights into precarious conditions that might be challenging to analyze otherwise. Simultaneously, the graph model serves as the actual DT, providing a versatile framework for running probabilistic simulations. This combination aims to enhance the capabilities of digital twins in healthcare, allowing for a more comprehensive and adaptable modeling of complex biological systems. However, a drawback of the aforementioned solution is its failure to address the need for a method facilitating the adaptation and generalization of the Patient DT within an MCPS. Altering a parameter in this context might demand intricate manual intervention on the GNN, potentially hindering the effective utilization of such a robust model. In our preliminary efforts, we tackle this limitation by ensuring the presence of a dedicated ontology-based module that interfaces with the more potent GNN. Any required modification will entail updating the structure and contents of the ontology. The ontology's syntax, grounded in concepts of classes and relations, is notably more user-friendly, even for individuals lacking expertise in the domain. Additionally, as an initial outcome, we present an implementation of our concept based on [61] within a Cloud-based framework.[3]

Design of a Biological Patient DT

We assume that the reader is familiar with the basic ML notions and models. However, we refer to [68] for further details. The proposed Patient DT consists of an AI-assisted modular system used to model the human body and to predict the evolution of patho-physiological conditions [61]. There are 2 main modules: *(i)* a GNN that provides clinically relevant endpoints, and *(ii)* a *generative adversarial network* (GAN) to generate synthetic data for simulating clinical situations.

Several properties of GNNs and GANs make them suitable for healthcare analysis.

- *non linearity*: both models are able to represent systems that are intrinsically non-linear in nature, such as pulmonary functions, heart rate and vascular dynamics.
- *interpretability*: graph-based models are much easier to interpret than other neural approaches as they do not require vectorization of the features of the input data; understanding models and their predictions is a critical aspect in healthcare.
- *non-euclidean geometry*: GNNs, founded on a non-Euclidean learning data structure, can be used to model biological systems at different scales, e.g. lymphatic vessels which can also be modeled as a graph where lymph nodes are the vertices.
- *modularity*: ability to learn independent modules that can be reused in different parts of the graph, in order to facilitate scalability.
- *cross-modality*: ability to combine structured and unstructured data in order to capture different levels of biological complexity.
- *generative*: ability to generate data preserving the statistical of the training set.
- *multiscale*: ability to integrate granular information at different levels of biological complexity, allowing the recognition of tissues, organs, processes, and systems.
- *spectral density*: by exploiting spatial properties, GNNs are amenable to analysis in the domain of frequency representation, enabling the analysis of motifs, substructures, and periodic patterns of the network.

Graph Neural Model

GNNs constitute a subset of deep learning models functioning within the graph domain, utilizing graph-based representations of studied samples as inputs. Operating as neural networks, GNNs acquire knowledge through a data-driven mechanism embedded in a multi-layer architecture, a crucial aspect for constructing hierarchical

[3] https://github.com/strumenti-formali-per-la-bioinformatica/patient-digital-twin.

models from data. GNNs, distinct from other deep learning models, harness additional features from graph theory and other mathematical theories.

The modular nature and capacity to devise intricate systems position GNNs as ideal for shaping a Patient DT, facilitating the stratification of human body layers. This involves *(i)* breaking down the complexity of the human body into independent genomic subsystems, capturing alterations, biological pathways, and organ physiology, *(ii)* presenting each derived subsystem as a node or network of nodes in a GNN, wherein interprocess signals undergo reformulation as message-passing operations, enabling the extraction of node embeddings, and *(iii)* ultimately, consolidating homogeneous subsystems into four layers, constituting the ultimate Patient DT model:

1. *transcriptomic*: acts on the set of RNA transcripts produced by the genome and on the gene expression analysis result which can be used to investigate the mechanism at the basis of a given disease [69–71].
2. *cellular*: involves biological processes that affect individual cells.
3. *organ*: includes organs and complex networks of cooperating organs (as proposed in [60], we have limited ourselves to the organs of the circulatory system and to a physiological representation of the heart, lungs, and kidneys).
4. *exposomic*: refers to deeply interconnected environmental factors that contribute to an individual's well-being or disease state (as proposed in [60], we consider *physical activity*, *dietary habits*, *viral infections*, and *therapeutic treatments*).

From a *"learning perspective"*, GNN solves the following possible levels of tasks: *(i) graph-level*, i.e., to predict a single property for the whole graph, *(ii) node-level* (resp. *edge-level*), i.e., to predict some property for each node (resp. for each edge).

To accomplish tasks such as learning embeddings for nodes, edges, and the global context of the graph during the training phase, GNNs leverage their structure organized in layers, typically consisting of multilayer perceptrons. This architecture allows the GNN to compute a transformation for each graph component. Here's a breakdown of the process:

1. *input vector*: vector of significant features representing a specific node or edge x.
2. *GNN Layer*: the GNN layer computes a *learned vector*, which serves as the embeddings representation for x; this construction of embeddings is achieved through the GNN layers using a technique known as *message passing*,
3. *message passing*: the *"neighborhood aggregation"* operation, where information for a node or edge is built by aggregating information from its neighbors; this operation is fundamental for the GNN to capture the relational dependencies within the graph.

By iteratively applying these layers and message passing operations, GNNs can capture complex relationships and dependencies within the graph, allowing them to learn meaningful embeddings for nodes, edges, and the overall graph structure. This process is crucial for the development of the Patient DT model [72].

Our proposed GNN aims to acquire the ability to accurately predict the clinical trajectory of a patient by forecasting clinically significant endpoints.

For this purpose, the GNN is constructed with nodes representing the variables found in the *ordinary differential equations* (ODEs) that describe the human body [60]. The edges represent the inherent relationships among these variables. Specifically, the computational framework outlined in [60] is employed to produce a time series of clinical endpoints for each differential equation, using a window size of $\tau = 500$ time steps. Given that the objective of each node is to grasp a latent representation, the strict mathematical structure of the ODE system is relaxed and learned directly from the data. Once trained and validated, the model is utilized to generate a set of potential trajectories.

The fundamental idea revolves around portraying a biological system within a GNN through the lens of differential equations. Initially, an ODE system is crafted from a biochemical reaction network. Following this, the ODE system is solved for diverse initial conditions, yielding a collection of trajectories for each variable. Ultimately, a GNN employs information from neighboring nodes to iteratively update the current state of the variable.

Generative Adversarial Model

GANs are generative models based on two-player gameplay where both players are encouraged to improve [67]. One player, the *generator*, creates samples that are intended to be indistinguishable from those from a given data distribution. The other player, the *discriminator*, learns to determine whether the samples come from the distribution (*real samples*) or from the real distribution of the fake data (*fake samples*).

The problem statement describes a dataset $\mathbf{D} = \{(x, m, r, q)\}$ consisting of samples from an unknown distribution $\mathbf{P}_{x,m,r,q}$. The components of each sample are defined as follows:

- $x \in \mathbf{R}^{t \times n}$: A matrix of n gene expression values in t tissues.
- $m \in \{0, 1\}^t$: a mask vector indicating whether the expression of each tissue has been measured for the given patient.
- $r \in \mathbf{R}^k$: a vector of k quantitative covariates (e.g., age).
- $q \in \mathbf{N}^c$: A vector of c categorical covariates (e.g., gender).

The goal is to generate realistic gene expression samples by modeling the conditional probability distribution $\mathbf{P}(X = x|M = m, R = r, Q = q)$. This modeling allows for sampling data under different conditions and assessing the uncertainty of the generated expression values.

To address this problem, a Wasserstein Generative Adversarial Network [73, 74] with gradient penalty (WGAN-GP) is proposed. WGAN-GP is an extension of Generative Adversarial Networks (GANs) that uses the Wasserstein distance and gradient penalty to stabilize and improve the training process. In the adversarial process, there are two key players:

1. The generator aims to produce samples from the conditional distribution $\mathbf{P}(X = x|M = m, R = r, Q = q)$.
2. The critic takes gene expression samples x from the generator and the data distribution, attempting to distinguish the true source of the input. The Wasserstein

distance is used to measure the dissimilarity between the generated distribution and the true data distribution.

The adversarial training process involves iteratively updating the generator and critic to improve the quality of generated samples while maintaining a stable training process. WGAN-GP introduces a gradient penalty to enforce a Lipschitz constraint on the critic, contributing to more stable and reliable training.

From Ontology to GNN

An *ontology* is a formal representation of knowledge within a specific domain. It defines concepts, classes, properties, and relationships in a structured manner to facilitate shared understanding and representation of knowledge. In the context of an ontology for a smart healthcare environment, these concepts are used to model various aspects of the smart healthcare domain, such as sensors, devices, medical conditions, etc. In our approach, the ontology serves a dual purpose: firstly, it has a "descriptive" objective, enabling the description of patient characteristics based on the "stratification of the human body" into layers. Secondly, it has a "computational" objective, providing the details necessary for the computational patient required in the construction of the previously described GNN. We have created an ontology called the *Patient DT ontology* that fulfills both criteria. We remark that, as proposed in [61], in this work, in order to demonstrate the flexibility and expressiveness of our GNN-based approach, we focus on modeling the *renin-angiotensin system* (`RAS`), i.e., the hormone system regulating vasoconstriction and inflammatory response, which represents one of the main biological pathways regulating blood pressure [75]. Our idea is to define one class for each variable used in the ODE system used to build the GNN, and one class for each biological layer of our patient DT described above. Furthermore, for each class corresponding to a biological layer we also explicitly specify the nodes involved and the relationships.

Now, we will provide an overview of the main classes and important properties in our ontology:

- *Classes*
  - `transcriptomic_layer`: this class elucidates the degree to which the expression of genes associated with the `RAS` can be clarified by genes originating from signaling and receptor pathways.
  - `cellular_layer`: this class delineates the biological processes that impact individual cells, encompassing metabolism, protein synthesis, replication, and motility. For instance, the `RAS` can be significantly influenced by factors such as glucose concentration, `ACE` inhibitor treatments, and viral infections that bind to `ACE2`, including SARS-CoV-2.
  - `organ_layer`: this class pertains to the circulatory system and provides a physiological representation of the heart, lungs, and kidneys.
  - `exposomic_layer`: this class refers to the overall exposure that individuals undergo from conception until death and its impact on diseases.

  - `edo_variable_node`: this class describes a variable involved in one of the differential equations of the ODE system, e.g. `ACE`, `ACE2`, `ANGI`, and `ANGII`.

- *Relationships*
  - `edge`: this link between two `edo_variable_node` objects represents the underlying relationships between the variables corresponding to such nodes.

Implementation Details

The implementation of our patient Digital Twin (DT) adhered to the *ML Operations* (MLOps) paradigm, a methodology designed to enhance the efficiency of developing and deploying machine learning (ML) models in production environments. MLOps encompasses tasks such as automating model training and deployment, managing data and computing resources, monitoring models in production, and facilitating continuous version updates.

The primary goal of MLOps is to streamline the life cycle of ML models, enabling organizations to capitalize on ML opportunities and promptly respond to market changes. Adoption of ML-specific best practices and tools, including model orchestration frameworks, version tracking systems, and model monitoring tools, is crucial for successful MLOps implementation. Several compelling reasons support the deployment of neural networks on the Cloud:

- *scalability*: Cloud services provide elastic computing and storage resources, allowing seamless scaling based on demand, a critical aspect for resource-intensive tasks like neural network training.
- *ease of use*: Many cloud providers offer fully managed ML services, eliminating concerns about managing underlying infrastructure.
- *accessibility*: Cloud computing and storage resources are accessible globally through the internet, benefitting organizations lacking powerful on-site hardware or those avoiding upfront hardware investments.
- *cost reduction*: Cloud usage entails paying only for consumed resources, sparing organizations from upfront hardware expenses and maintenance costs.
- *availability*: Cloud providers ensure high availability and redundancy, reducing the likelihood of service interruptions due to hardware issues.

The proposed architecture involves creating a Docker image containing all necessary dependencies for launching pre-existing Python scripts, sourced from [61], in a portable and secure manner. Key dependencies on the Docker image include PyTorch and NumPy. Pre-installed dependencies on the `docker`[4] image include `pytorch`[5] and `numpy`.[6] The Docker image is generated every time a repository contributor pushes via GitHub. The push trigger allows you to initialize a `CI/CD` routine implemented via `GitHub Actions` (see Fig. 10).

---

[4] https://www.docker.com/.

[5] https://pytorch.org/.

[6] https://numpy.org/.

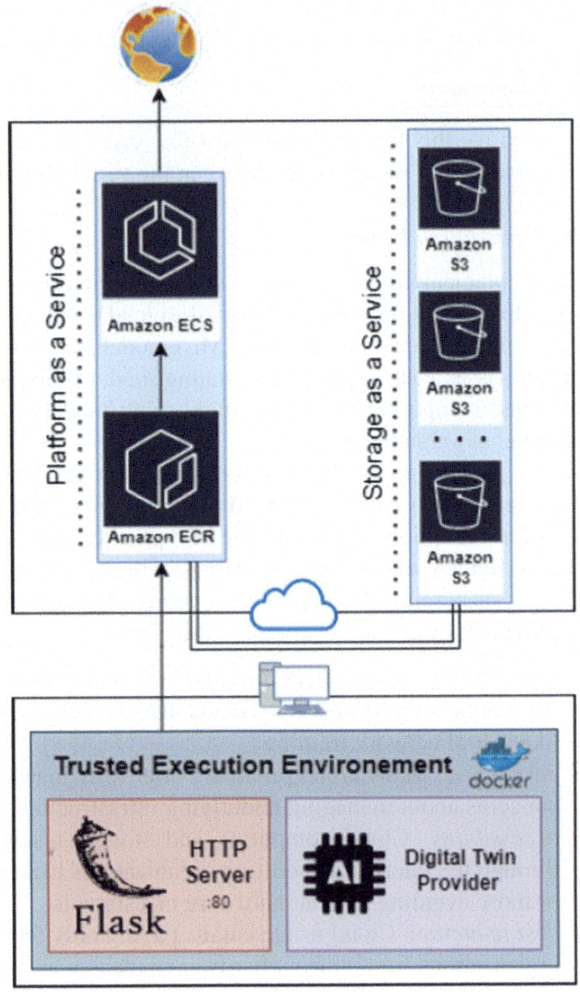

Fig. 10 Proposed hardware architecture for *Patient DT* with *MLOps* methodologies

`Continuous Integration/Continuous Deployment` (CI/CD) is a software development methodology focused on automating the integration and deployment of an application's source code. In practical terms, the `CI/CD` process entails automatic compilation, testing, and deployment whenever a developer commits their code, eliminating the need for manual intervention. The primary advantage of `CI/CD` lies in its ability to reduce the release time of an application, as code is automatically tested and deployed, eliminating the waiting period for manual operations by developers. Moreover, `CI/CD` contributes to maintaining code quality by automating the integration and deployment processes, ensuring frequent repetitions that allow prompt identification and resolution of issues. Various workflow automa-

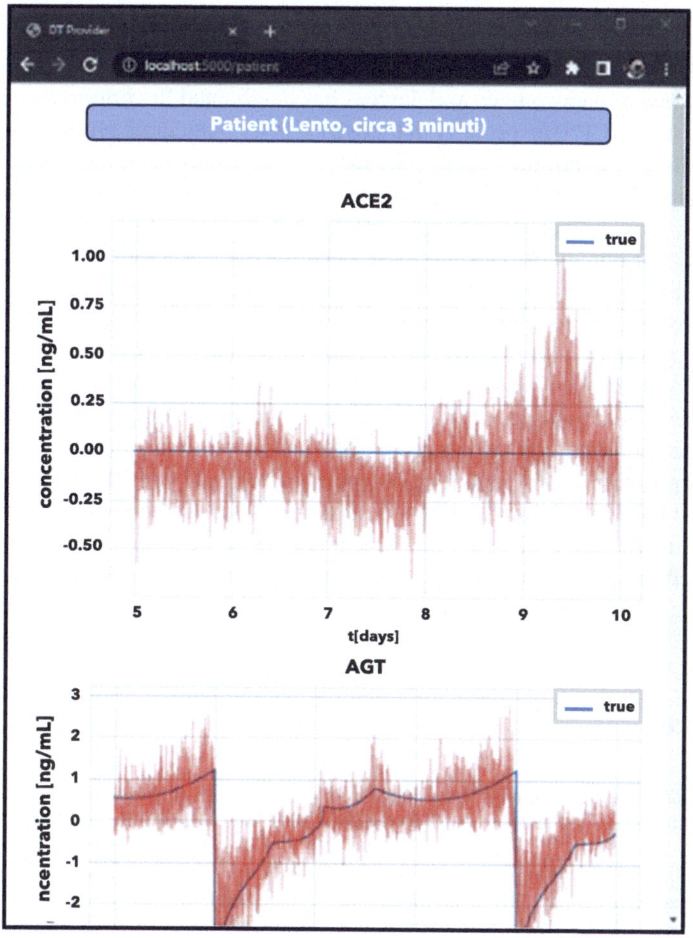

Fig. 11 Application Front-end developed to visualize the prediction provided by the Patient DT

tion platforms, including `GitHub Actions`, are available for streamlining and automating the `CI/CD` pipeline.

The application's front-end is designed to enable minimal and restricted interaction with the Digital Twin Provider. It is deployed within a self-contained Docker container, ensuring isolation from the underlying operating system. This virtualized environment allows the installation of essential Python packages, like `Flask`, crucial for back-end operations, without affecting the host machine and avoiding potential versioning conflicts (see Fig. 11).

In general, the platform, created as a proof of concept, allows you to connect to the provider via a bidirectional channel created via polling. When pressing a specific button, it will come as a `RESTful API` is invoked (organically indicated among the routes of the `Flask` webapp) which, in turn, will forward the request to the actual digital twin provider, that finally will transmit the results.

4 Conclusions

The global rise of data sources and devices has heightened the importance of cyber-physical systems (CPS), especially in Industry 4.0. CPS integrate physical and virtual processes through continuous feedback loops, monitored by computers and networks. In healthcare, CPS enable real-time monitoring of patient lifestyles, diets, and health parameters, providing alerts for prescriptions, aiding medication selection, and predicting surgical outcomes. Despite these benefits, there is a lack of comprehensive guidelines for maximizing CPS's potential in healthcare, highlighting the need for advanced techniques to enhance patient care and drive digital transformation.

This study aims to address this gap by elucidating and organizing guidelines, with a specific focus on clarifying distinctions between CPS and Digital Twins (DT), two interconnected technologies with significant implications for healthcare. As part of our proposed solution, we introduce an innovative medical model, leveraging Geometric Deep Learning for patient representation (referred to as the Patient DT) and employing Ontologies to streamline the translation of intricate healthcare data into a graph-based structure utilized by the Patient DT. This integrative approach promises enhanced efficiency and adaptability in modeling complex biological systems within the healthcare domain.

References

1. Abujassar RS (2020) A new smart algorithm to provide an integrity path in networks for improving the abroad healthcare surgery setting. Int J Innov Comput Inf Control 16(3):1047–1058
2. Assembly WHO (2005) World health organization assembly. https://www.who.int/healthacademy/media/WHA58-28-en.pdf?ua=1. Accessed Feb 2021
3. Lee J, Bagheri B, Kao H-A (2015) A cyber-physical systems architecture for industry 4.0-based manufacturing systems. Manuf Lett 3:18–23
4. Gupta V, Singh Gill H, Singh P, Kaur R (2018) An energy efficient fog-cloud based architecture for healthcare. J Stat Manag Syst 21(4):529–37
5. Aceto G, Persico V, Pescapé A (2020) Industry 4.0 and health: internet of things, big data, and cloud computing for healthcare 4.0. J Ind Inf Integr. 18:100129
6. Rao CS, Shankar RS, Murthy K (2019) Cyber-physical systems-an overview. In: Smart intelligent computing and applications, pp 489–497
7. Humayed A, Lin J, Li F, Luo B (2017) Cyber-physical systems security-a survey. IEEE Internet Things J 4(6):1802–31
8. Munir A, Kansakar P, Khan SU (2017) IFCIoT: integrated fog cloud IoT: a novel architectural paradigm for the future internet of things. IEEE Consum Electron Mag 6(3):74–82
9. Chaâri R, Ellouze F, Koubâa A, Qureshi B, Pereira N, Youssef H, Tovar E (2016) Cyber-physical systems clouds: a survey. Comput Netw 108:260–78
10. Tortorella GL, Fogliatto FS, Mac Cawley Vergara A, Vassolo R, Sawhney R (2020) Healthcare 4.0: trends, challenges and research directions. Prod Plan Control 31(15):1245–1260
11. Zhang Y, Qiu M, Tsai C-W, Hassan MM, Alamri A (2015) Health-CPS: healthcare cyber-physical system assisted by cloud and big data. IEEE Syst J 11(1):88–95
12. Abdellatif AA, Mohamed A, Chiasserini CF, Tlili M, Erbad A (2019) Edge computing for smart health: context-aware approaches, opportunities, and challenges. IEEE Netw 33(3):196–203

13. Mowla N, Doh I, Chae K (2018) Selective fuzzy ensemble learner for cognitive detection of bio-identifiable modality spoofing in MCPS. In: Proceeding of the 20th international conference on advanced communication technology (ICACT), pp 63–67
14. Alabdulatif A, Khalil I, Yi X, Guizani M (2019) Secure edge of things for smart healthcare surveillance framework. IEEE Access 7:31010–31021
15. Roehrs A, da Costa CA, da Rosa Righi R, da Silva VF, Goldim JR, Schmidt DC (2019) Analyzing the performance of a blockchain-based personal health record implementation. J Biomed Inform 92, Art. no. 103140
16. Verma P, Sood SK (2018) Fog assisted-IoT enabled patient health monitoring in smart homes. IEEE Internet Things J 5(3):1789–96
17. Raghupathi W, Raghupathi V (2014) Big data analytics in healthcare: promise and potential. Health Inf Sci Syst 2(1):1–10
18. Anuradha J et al (2015) A brief introduction on big data 5 vs characteristics and Hadoop technology. Proc Comput Sci 48:319–24
19. Banerjee A, Venkatasubramanian KK, Mukherjee T, Gupta SKS (2011) Ensuring safety, security, and sustainability of mission-critical cyber-physical systems. Proc IEEE 100(1):283–299
20. Murugesan A et al (2014) From requirements to code: model based development of a medical cyber physical system. In: Proceedings of the international workshop on software engineering for health care, pp 96–112
21. Wu Z et al (2016) A community health service architecture based on the Internet of things on health-care. Chin J Biomed Eng 25(3):114–120
22. Xu E, Wermus M, Bauman DB (2011) Development of an integrated medical supply information system. Enterprise Inf Syst 5(3):385–399
23. Din S, Paul A (2019) Smart health monitoring and management system: toward autonomous wearable sensing for internet of things using big data analytics. Future Gener Comput Syst 91:611–619
24. Ma B et al (2020) Assuring privacy-preservation in mining medical text materials for COVID-19 cases-a natural language processing perspective. In: Proceedings of the VLDB international workshop very large internet things (VLIoT), pp 6–13
25. Jiemin Z, Jinsheng L (2010) The district medical data center based on cloud computing. In: Proceedings of the 5th international conference on computer science & education, pp 1424–1427
26. Xue T et al (2017) Research on medical data sharing model based on blockchain. Acta Autom Sin 43(9):1555–1562
27. Grieves M (2014) Digital twin: manufacturing excellence through virtual factory replication. White paper 1(2014):1–7
28. Liu Y, Zhang L, Yang Y, Zhou L, Ren L, Wang F, Deen MJ (2019) A novel cloudbased framework for the elderly healthcare services using digital twin. IEEE Access 7:49088–49101
29. Elayan H, Aloqaily M, Guizani M (2021) Digital twin for intelligent context-aware IoT healthcare systems. IEEE Internet Things J 8(23):16749–16757
30. Erol T, Mendi AF, Dogan D (2020) The digital twin revolution in healthcare. In: 2020 4th international symposium on multidisciplinary studies and innovative technologies (ISMSIT). IEEE, pp 1–7
31. Haleem A, Javaid M, Singh RP, Suman R (2022) Medical 4.0 technologies for healthcare: features, capabilities, and applications. Internet Things Cyber Phys Syst
32. Feng Y, Chen X, Zhao J (2018) Create the individualized digital twin for noninvasive precise pulmonary healthcare. Significances Bioeng & Biosci 1(2)
33. Fuller A, Fan Z, Day C, Barlow C (2020) Digital twin: enabling technologies, challenges and open research. IEEE Access 8:108952–108971
34. Rivera LF, Jimenez M, Angara P, Villegas NM, Tamura G, Müller HA (2019) Towards continuous monitoring in personalized healthcare through digital twins. In: Proceedings of the 29th annual international conference on computer science and software engineering, pp 329–335
35. Ahmed H, Devoto L (2021) The potential of a digital twin in surgery. Surg Innovat 28(4):509–510

36. Javaid M, Haleem A, Singh RP, Khan S, Suman R (2021) Blockchain technology applications for industry 4.0: a literature-based review. Blockchain Res Appl 100027
37. Barricelli BR, Casiraghi E, Fogli D (2019) A survey on digital twin: definitions, characteristics, applications, and design implications. IEEE Access 7:167653–167671
38. Popa EO, van Hilten M, Oosterkamp E, Bogaardt MJ (2021) The use of digital twins in healthcare: socio-ethical benefits and socio-ethical risks. Life Sci Soc Policy 17(1):1–25
39. De Benedictis A, Mazzocca N, Somma A, Strigaro C (2022) Digital twins in healthcare: an architectural proposal and its application in a social distancing case study. IEEE J Biomed Health Inform. https://doi.org/10.1109/JBHI.2022.3205506
40. Okegbile SD, Cai J, Yi C, Niyato D (2022) Human digital twin for personalized healthcare: vision. Architecture and future directions. IEEE Netw
41. El Marai O, Taleb T, Song J (2020) Roads infrastructure digital twin: a step toward smarter cities realization. IEEE Netw 35(2):136–143
42. Heininger R, Stary C (2021) Capturing autonomy in its multiple facets: a digital twin approach. In: Proceedings of the 2021 ACM workshop on secure and trustworthy cyber-physical systems, pp 3–12
43. Liu YK, Ong SK, Nee AYC (2022) State-of-the-art survey on digital twin implementations. Adv Manuf 10(1):1–23
44. Madubuike OC, Anumba CJ (2021) Digital twin application in healthcare facilities management. Comput Civ Eng 366–373
45. Song Y, Li Y (2022) Digital twin aided healthcare facility management: a case study of Shanghai tongji hospital. In: Construction research congress, pp 1145–1155
46. Hu L, Xie N, Kuang Z, Zhao K (2012) Review of cyber-physical system architecture. In: Proceedings of the 2012 15th IEEE international symposium on object/component/service-oriented real-time distributed computing workshops. ISORCW, pp 25–30
47. Liu Y, Peng Y, Wang B, Yao S, Liu Z (2017) Review on cyber-physical systems. IEEE/CAA J Autom Sin 4(1):27–40
48. Win SY, Thianwiboon M (2021) Parametric optimization of NACA 4412 airfoil in ground effect using full factorial design of experiment. Eng J 25(12):9–19
49. Parnianifard A, Rezaie V, Chaudhary S, Imran MA, Wuttisittikulkij L (2021) New adaptive surrogate-based approach combined swarm optimizer assisted less tuning cost of dynamic production- inventory control system. IEEE Access 9:144054–144066
50. Amaruchkul K, Pongsathornwiwat A, Bantadtiang P (2022) Constrained joint replenishment problem with refrigerated vehicles. Eng J 26(1):75–91
51. Wang H, Deng X, Tian F (2012) WiP abstract: a human-centered cyber-physical systematic approach for post-stroke monitoring. In: 2012 IEEE/ACM third international conference on cyber-physical systems, p 209
52. Ghorbani M, Bogdan P (2013) A cyber-physical system approach to artificial pancreas design. In: 2013 international conference on hardware/software codesign and system synthesis, CODES+ISSS, pp 1–10
53. Bécue A, Fourastier Y, Praça I, Savarit A, Baron C, Gradussofs B et al (2018) Cyberfactory#1—securing the industry 4.0 with cyber-ranges and digital twins. In: 2018 14th IEEE international workshop on factory communication systems (WFCS), pp 1–4
54. Bitton R, Gluck T, Stan O, Inokuchi M, Ohta Y, Yamada Y et al (2018) Deriving a cost-effective digital twin of an ICS to facilitate security evaluation. In: Lopez J, Zhou J, Soriano M (eds) Computer security. Springer, Cham, pp 533–554
55. Eckhart M, Ekelhart A (2018c) Towards security-aware virtual environments for digital twins. In: Proceedings of the 4th ACM workshop on cyber-physical system security, CPSS '18, pp 61–72. ACM, New York, NY
56. Tauber M, Schmittner C (2018) Enabling security and safety evaluation in industry 4.0 use cases with digital twins. ERCIM News 2018(115)
57. Axelrod CW (2013) Managing the risks of cyber-physical systems. In: 2013 IEEE long island systems. Applications and technology conference (LISAT), pp 1–6

58. Venkatasubramanian KK, Nabar S, Gupta SKS, Poovendran R (2012) Cyber physical security solutions for pervasive health monitoring systems. In: E-healthcare systems and wireless communications: current and future challenges. IGI Global, pp 143–162
59. Arney D, Pajic M, Goldman JM, Lee I, Mangharam R, Sokolsky O (2010) Toward patient safety in closed-loop medical device systems. In: Proceedings of the 1st ACM/IEEE international conference on cyber-physical systems, pp 139–148
60. Barbiero P, Lió P (2020) The computational patient has diabetes and a covid. arXiv:2006.06435
61. Barbiero P, Vinas Torne R, Lió P (2021) Graph representation forecasting of patient's medical conditions: toward a digital twin. Front Genet 12:652907
62. Zitnik M, Agrawal M, Leskovec J (2018) Modeling polypharmacy side effects with graph convolutional networks. Bioinformatics 34:i457–i466. https://doi.org/10.1093/bioinformatics/bty294.
63. Ginsburg GS, Willard HF (2009) Genomic and personalized medicine: foundations and applications. Transl Res 154:277–287. https://doi.org/10.1016/j.trsl.2009.09.005.
64. Naylor S, Chen JY (2010) Unraveling human complexity and disease with systems biology and personalized medicine. Pers Med 7:275–289. https://doi.org/10.2217/pme.10.16.
65. Laubenbacher R, Sluka JP, Glazier JA (2021) Using digital twins in viral infection. Science 371:1105–1106. https://doi.org/10.1126/science.abf3370
66. Scarselli F, Gori M, Tsoi AC, Hagenbuchner M, Monfardini G (2008) The graph neural network model. IEEE Trans Neural Netw 20:61–80. https://doi.org/10.1109/TNN.2008.2005605
67. Goodfellow I, Pouget-Abadie J, Mirza M, Xu B, Warde-Farley D, Ozair S et al (2014) Generative adversarial nets. Adv Neural Inf Syst Process Syst 2672–2680
68. Yang J, Li Y, Liu Q, Li L, Feng A, Wang T, Lyu J (2020) Brief introduction of medical database and data mining technology in big data era. J Evid Based Med 13(1):57–69
69. Emilsson V, Thorleifsson G, Zhang B, Leonardson AS, Zink F, Zhu J et al (2008) Genetics of gene expression and its effect on disease. Nature 452:423–428. https://doi.org/10.1038/nature06758
70. Cookson W, Liang L, Abecasis G, Moffatt M, Lathrop M (2009) Mapping complex disease traits with global gene expression. Nat Rev Genet 10:184–194. https://doi.org/10.1038/nrg2537
71. Gamazon, E. R., Segrè, A. V., van de Bunt, M., Wen, X., Xi, H. S., Hormozdiari, F., et al. (2018). Using an atlas of gene regulation across 44 human tissues to inform complex disease- and trait-associated variation. 50, 956-967. Nat. Genet. doi: https://doi.org/10.1038/s41588-018-0154-4
72. Rocco Zaccagnino, Antonio Cirillo, Alfonso Guarino, Nicola Lettieri, Delfina Malandrino, Gianluca Zaccagnino: Towards a Geometric Deep Learning-Based Cyber Security: Network System Intrusion Detection Using Graph Neural Networks. SECRYPT 2023: 394-401
73. Arjovsky, M., Chintala, S., and Bottou, L. (2017). Wasserstein GAN. arXiv [Preprint]. arXiv:1701.07875
74. Gulrajani, I., Ahmed, F., Arjovsky, M., Dumoulin, V., and Courville, A. (2017). Improved training ofWasserstein GANs. arXiv [Preprint]. arXiv: 1704.00028
75. Fountain, J. H., and Lappin, S. L. (2019). Physiology, renin angiotensin system, in StatPearls, (StatPearls Publishing)
76. Biolchini, J., Mian, P., Natali, A. & Travassos, G. Systematic review in software engineering. *System Engineering And Computer Science Department COPPE/UFRJ, Technical Report ES.* **679**, 45 (2005)
77. Kitchenham, B., Charters, S. & Others Guidelines for performing systematic literature reviews in software engineering. (UK,2007)
78. Roehrs A, Da Costa C, Rosa Righi R, De Oliveira K (2017) & Others Personal health records: a systematic literature review. Journal Of Medical Internet Research. 19:e5876
79. Petticrew, M. & Roberts, H. Systematic reviews in the social sciences: A practical guide. (John Wiley & Sons,2008)
80. Aria M, Cuccurullo C (2017) bibliometrix: An R-tool for comprehensive science mapping analysis. Journal Of Informetrics. 11:959–975

81. Gobbo, E., Guarino, A., Cafarelli, B., Grilli, L. & Limone, P. Automatic evaluation of open-ended questions for online learning. A systematic mapping. *Studies In Educational Evaluation*. **77** pp. 101258 (2023)
82. Cobo M, López-Herrera A, Herrera-Viedma E, Herrera F (2011) An approach for detecting, quantifying, and visualizing the evolution of a research field: A practical application to the Fuzzy Sets Theory field. Journal Of Informetrics. 5:146–166
83. Callon M, Courtial J, Laville F (1991) Co-word analysis as a tool for describing the network of interactions between basic and technological research: The case of polymer chemsitry. Scientometrics. 22:155–205
84. Aria M, Cuccurullo C, D'Aniello L, Misuraca M, Spano M (2022) Thematic analysis as a new culturomic tool: the social media coverage on COVID-19 pandemic in Italy. Sustainability. 14:3643
85. Garg H, Sharma B, Shekhar S, Agarwal R (2022) Spoofing detection system for e-health digital twin using EfficientNet Convolution Neural Network. Multimedia Tools And Applications. 81:26873–26888

Federated Learning in Agents Based Cyber-Physical Systems

Domenico Di Sivo, Palma Errico, and Salvatore Venticinque

Abstract The pervasiveness of connected wearable devices, combined with recent advances in Artificial Intelligence (AI), which is capable of extracting significant insights from vast amounts of collected data, has driven the development of cyber-physical systems composed of human users and intelligent agents. Multi-agent systems (MAS) represent a powerful paradigm to develop such kind of distributed systems. They can autonomously learn from data and assist users to manage personal information, supporting individuals in their daily activities. In the smart-health context, sharing personal health data with physicians can create information flows that enhance the processes of diagnosis, monitoring, and treatment for a growing number of conditions, particularly in frail patients. However, with increasing emphasis on data privacy and security, there is heightened awareness of the risks associated with collecting data from all users for machine learning tasks. Federated Learning (FL) has emerged as a solution, offering a decentralized framework where a shared prediction model can be built while ensuring that owners' data remains on their own devices. In FL, the data residing on various devices may exhibit imbalances and non-i.i.d. (independent and identically distributed) characteristics. In this chapter, we introduce the main concepts of the FL paradigm and the relevance of MAS as an enabling technology to exploit FL in decentralized environments such as edge devices or IoT networks.

Domenico Di Sivo is a PhD student enrolled in the National PhD in Artificial Intelligence, XXXVIII cycle, course on Health and life sciences, organized by "Universitá Campus Bio-Medico di Roma".

D. Di Sivo · P. Errico · S. Venticinque (✉)
University of Campania Luigi Vanvitelli, Aversa, (CE), Italy
e-mail: salvatore.venticinque@unicampania.it

P. Errico
e-mail: palma.errico@unicampania.it

M. Ficco and G. D'Angelo (eds.), *Artificial Intelligence Techniques for Analysing Sensitive Data in Medical Cyber-Physical Systems*, Engineering Cyber-Physical Systems and Critical Infrastructures 12, https://doi.org/10.1007/978-3-031-70775-9_5

1 Introduction

As the Internet of Things (IoT) continues to proliferate, with projections suggesting that half of all devices will be IoT objects in the near future, according to Cisco's latest report[16], there is a growing emphasis on leveraging IoT solutions for various applications. The proliferation of mobile devices capable of collecting diverse types of data, including sensory data like GPS coordinates, accelerometers, and various sensor data, as well as images, has seen a significant surge. Traditional IoT architectures typically involve interconnected devices, with lower-level nodes responsible for data acquisition and higher-level nodes handling processing and analytics tasks.

Driven by the abundance of big data, artificial intelligence has permeated various spheres of daily life, including transportation, agriculture, insurance, and healthcare. However, the conventional approach to AI necessitates amalgamating data from disparate sources, which can be costly and time-consuming.

The last advances in Artificial Intelligence (AI), which is able to extract more and more knowledge from this huge amount of collected data, leveraged the development of cyber-physical systems composed of human users and intelligent agents. In this centralized architecture, Intelligent Agents can extract knowledge from a theoretically infinite amount of data and exploit their context awareness to support human users in their daily activities.

However, stringent data privacy regulations, such as the GDPR in the EU and the General Security Law in China, impose restrictions on data sharing among organizations. With the increasing emphasis on data privacy and security, there is a growing awareness of the risks associated with collecting data from all users for machine learning tasks. Federated learning emerges as a solution to this challenge, offering a decentralized framework where a shared prediction model can be built, distributed learners without compromising data privacy, ensuring that owners' data remains on their own devices.

Initially designed by Google for mobile and edge computing, federated learning has expanded to encompass secure distributed learning across diverse organizations, fostering collaboration while preserving data privacy. Recognizing the potential of federated learning to facilitate data collaboration, the federated machine learning group has initiated work on an IEEE standard (IEEE Std 3652.1-2020) for federated learning, aimed at reducing the costs and risks associated with AI projects involving data from multiple sources.

Moreover, several challenges, including higher data transfer volumes, longer processing times, and data security and privacy concerns, must be addressed. An further significant challenge facing this technique is achieving parallelization across all IoT devices [51], taking into account their dynamic heterogeneity in terms of hardware/software characteristics, utilization and data collection.

The exploitation of collaborative agents to build collective intelligence has been demonstrated to be effective in various research works [32, 49]. Social ability and pro-activeness, which characterize Multi-agent system (MAS) technology, can be exploited to improve necessary data exchange and the aggregation process in a federated learning system by improving its flexibility and dynamism [43].

In the first part of this chapter we introduce application scenarios in the context of e-health. Then, we discuss different types of federated learning approaches, considering factors such as data distribution and communication patterns. The major concerns related to federated learning, including data privacy, security, and model performance, will also be addressed. Therefore the second part of the chapter provides a specific focus on the exploitation of the MAS programming paradigm and technology for enhancing the development and deployment of FL solutions.

2 Application Scenarios

In this section, we present three examples of application scenarios where human users and intelligent agents collaborate to utilize collected data to enhance human well-being.

2.1 *Collaborative Symptoms Tracking and Forecasting*

Smart apps increasingly log information such as diet and symptoms, often integrating data from mobile phones, smartwatches, or fitness trackers. In this context, applications act as personal assistants, analyzing user data, presenting statistics, and comparing results with data from other users who share their information. By pooling this data, these apps can learn from a centralized repository, enabling better classification of individual situations using statistical analysis and trained models for classification or prediction. This also facilitates community-level investigations into emerging behaviors, allowing for the monitoring of symptom patterns and the prevention of potential epidemics. Additionally, it helps track the progression of known diseases and identify significant changes.

Typically, these apps collect data and train statistical or machine learning models in the cloud, executing them remotely. However, federated learning (FL) could change this approach by sharing only aggregated statistics or partially trained models, which are processed locally and then combined centrally. This method ensures user privacy while enabling advanced collaboration between centralized systems and individual apps. It also fosters a new level of security and cooperation, allowing personal assistants to train, exchange, and update models while safeguarding user data.

2.2 *Correlate Health and Disease Condition*

Correlating health conditions and diseases is fundamental in modern medicine, as it offers the potential to identify risk factors and develop preventive strategies. Imagine someone falling ill; for a doctor, it would be extremely valuable to have information about the person's health status before the onset of the illness, even if this data is not derived from traditional medical sources. For example, data on sleep, physical activity, diet, and stress can provide crucial clues about conditions that might predispose someone to illness.

This type of data would be even more useful if collected from a broader population, including both people who have fallen ill and those who haven't. This approach would enable comparisons of lifestyles, identification of early symptoms, and observation of behavioral patterns that could indicate a predisposition to illness. The availability of such data could pave the way for new approaches to prevention, early diagnosis, and treatment.

The adoption of federated learning techniques would increase the level of acceptance by those people which are less sensitive to the problem and concerned about their privacy. Their participation would make it possible to gain a more detailed understanding of the correlations between health conditions and diseases.

In this scenario, a personal agent could play a key role in collecting a set of personal data and managing this information. The agent could monitor aspects such as physical activity, diet, sleep, and other daily behaviors, helping the user to create a complete profile of their health. Additionally, the agent could suggest that the user voluntarily participate in a training and testing campaign, where the collected data is used to fuel machine learning models designed to predict and prevent diseases.

The voluntary aspect is crucial in this context, as it ensures respect for privacy and informed consent. Moreover, a personal agent should be designed to ensure data security and anonymity, sharing only aggregated information or statistics, thus avoiding revealing personal details that could compromise the user's privacy.

2.3 Multi-modal Monitoring

When a disease has already manifested, clinical monitoring requires specific medical devices and often involves challenges for the patient entailing logistical hurdles, financial constraints, and potential physical discomfort. Non-medical devices, such as wearable technology, offer a less invasive means of collecting data and extracting general information about a person's well-being between medical visits. This information can complement the clinical framework.

A personal agent, equipped with models trained on clinical monitoring results to correlate medical data with non-medical data, could support both patients and doctors' decisions, such as advancing clinical monitoring or providing insights into the impact of diseases on quality of life. In a broader context, this approach could contribute to identifying emerging patterns in the overall well-being of a community and predicting the spread of an epidemic in advance.

As an example, polysomnography is one of the most comprehensive methods for diagnosing sleep disorders, but it is also notoriously invasive due to the procedure and the equipment used. Moreover, during a polysomnography test, the patient must sleep in a controlled environment, often a sleep lab, with numerous sensors that monitor physiological functions such as brain activity, eye movements, muscle activity, respiratory movements, airflow, blood oxygen levels, and heart activity.

A non medical ballistocardiographic (BCG) sensors can capture heartbeats, respiration, and body movements, which can be easily used by patients at home. It cannot be used for sleep disorder diagnosis, but a multi-modal learning which is trained with both devices, at home and in the controlled environment, could be used to extend the awareness and the capability to infer relevant information about the user's health.

3 Distributed Learning

Distributed learning embodies the convergence of computational power and algorithmic intelligence by trough an amalgamation of distributed computing and machine learning. Its overarching objective is to orchestrate a comprehensive estimation by harnessing the collective capabilities of distributed computing units and aggregating their results. Federated learning, a specialized form of distributed learning, shares this distributed approach, but diverges in crucial aspects from traditional distributed learning paradigms.

Unlike conventional distributed learning frameworks, federated learning demands feedback on the global model's performance from participating devices or nodes. This feedback loop enables the refinement and enhancement of the global model iteratively, leveraging insights gained from diverse data sources. Moreover, federated learning places a heightened emphasis on security and privacy considerations due to the decentralized nature of data storage and processing. Unlike centralized learning systems, where data is aggregated in one location, federated learning operates on distributed data sources, necessitating robust mechanisms for safeguarding sensitive information.

These distinctions between federated learning and conventional distributed learning introduce a range of novel challenges. The decentralized nature of federated learning complicates model coordination and synchronization across heterogeneous devices and networks. Moreover, federated learning must address concerns related to data heterogeneity, bias, and representativeness to ensure the robustness and generalization capability of the learned models.

In summary, while distributed learning serves as a powerful paradigm for harnessing collective computing resources, federated learning introduces unique complexities and considerations, particularly concerning model feedback, security, and privacy, necessitating innovative solutions and methodologies to address these challenges effectively.

Multi-agent systems are a particular case of a distributed system, and its particularity lies in the fact that the components of the system are autonomous and selfish, seeking to satisfy their own objectives. In addition, these systems also stand out for being open systems without a centralized design [12]. One main reason for the great interest and attention that multi-agent systems have received is that they are seen as an enabling technology for complex applications that require distributed and parallel processing of data and operate autonomously in complex and dynamic domains.

Learning in Multi-Agent Systems (MAS) is a crucial paradigm, as a system that can learn and dynamically adapt its actions offers significant potential to address the challenges of federated learning. A unique characteristic of MAS is that agents do not operate in isolation; they must consider other agents who also perceive and act within the environment. MAS provides essential mechanisms such as discovery, presence, and interaction protocols, including negotiation and brokering.

4 Federated Learning

Federated machine learning represents a distributed framework wherein dataset owners collectively train a global model for specific tasks, such as classification, prediction, or regression. Unlike traditional methods, federated learning enables collaborative model training without the need to exchange raw data, ensuring the privacy of each participant. The primary objective of federated learning is to achieve performance comparable to that of centralized approaches, while maintaining data privacy. A practical example is presented in [28]. In this scenario, user devices interact with a cloud server periodically to conduct distributed training for next word prediction on mobile phones. During each communication round, a subset of mobile users undertake local training using their respective datasets. Instead of transmitting raw data, the devices upload trained model parameters to the cloud server. After aggregating these updates, the central server redistributes the refined global model to another subset of users, continuing this iterative process until a predefined stopping criterion is reached. The experiment detailed in [28] demonstrates the efficacy of federated learning in producing high-quality models for next word prediction while safeguarding user data privacy at the local level. Figure 1 depicts a federated learning framework for an e-health infrastructure where, instead of transmitting raw data to a central server, remote devices periodically communicate with it to train a global model. During each communication round, a subset of devices conducts local training on their user data and sends updates to the server. The server integrates these updates and redistributes the updated global model to other devices. This iterative process continues until convergence or a stopping criterion is met. Alternative architectures, which introduce aggregators or hierarchical layers, have been proposed in [8] and in [52] to limit successful inference and reconstruction attacks.

4.1 Application Domains

While ongoing research continues to refine its methodologies, federated learning has already found practical applications across diverse domains. The *IEEE Std 3652.1-2020* standard delineates the diverse application domains to which federated learning can be applied, elucidating the contexts in which each category finds relevance and utility.

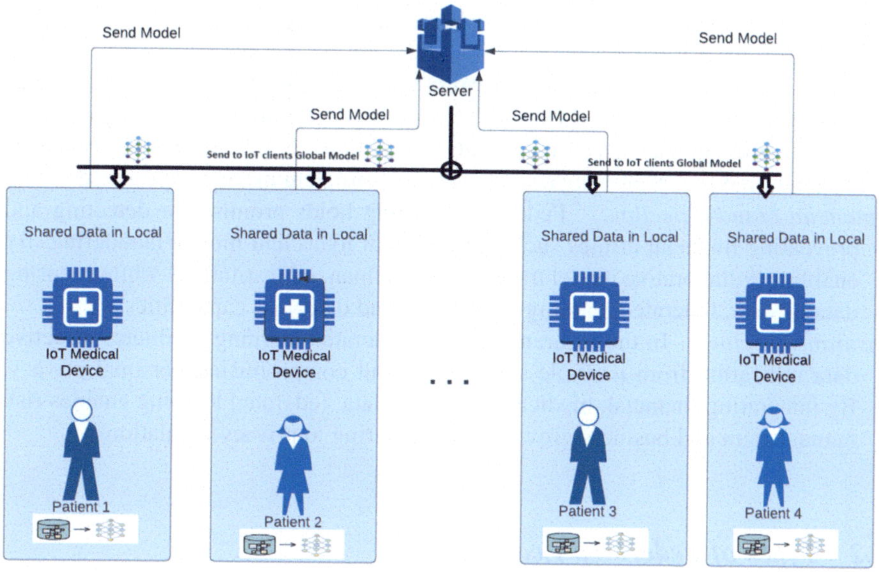

Fig. 1 An illustrative representation of the federated learning framework

Smartphone applications: Utilizing statistical models trained on aggregated user behavior data from a diverse array of mobile devices, federated learning powers various smartphone applications such as next-word prediction, facial recognition, and voice recognition. This ensures accurate predictions while safeguarding user privacy and minimizing resource consumption.

Organizational Data: In federated learning, entire organizations or institutions can be treated as *devices*. For instance, hospitals possess vast amounts of patient data crucial for predictive healthcare applications. Federated learning enables collaborative model training while respecting privacy and reducing network loads, making it an ideal solution for organizations with stringent privacy policies.

Internet of Things (IoT): Federated learning facilitates the development of models that dynamically respond to changes in IoT networks like wearable devices, autonomous cars, and smart homes. By preserving user privacy and addressing connectivity limitations, federated learning ensures efficient learning from distributed data sources.

Healthcare Industry: Given the stringent regulations surrounding the sharing of sensitive health information, federated learning offers significant benefits to the healthcare sector. It enables collaborative AI model training across various healthcare databases and devices while ensuring compliance with regulatory standards.

Personalized Advertising: Amid growing concerns about data privacy, federated learning provides a viable solution for personalized advertising on platforms like social media and e-commerce websites. By preserving user privacy preferences, federated learning enables personalized experiences without compromising individual privacy.

Autonomous Vehicles: Federated learning contributes to the development of real-time predictive capabilities essential for safe navigation in self-driving cars. By incorporating updates on road and traffic conditions, federated learning enhances decision-making, promising a safer and more efficient driving experience.

Financial Fraud Detection: Federated learning holds promise for detecting and preventing financial crimes such as fraudulent loans and money laundering. By enabling collaborative model training across financial institutions while ensuring data privacy, federated learning enhances fraud detection capabilities.

Insurance Sector: In the insurance sector, federated learning facilitates effective data utilization from multiple sources without compromising personal privacy. By integrating financial, medical, and other data, federated learning enables risk management and business growth while adhering to privacy regulations.

4.2 Types of Federated Learning

In federated machine learning, data is typically stored in a standardized database format, often resembling a table structure. Let's envision a scenario where a data owner possesses a dataset comprising numerous data samples. Each of these data samples includes both features (the input variables) and labels (the corresponding output or target variable), and each sample is assigned a unique identifier, known as a sample ID. In federated learning, datasets from various owners may share sample IDs and/or feature attributes. The nature and extent of this overlap determine the classification of federated learning into different categories.

Horizontal federated learning occurs when datasets from different owners share common sample IDs, meaning they have overlapping data instances. In this scenario, the data samples from different sources possess similar feature attributes, but might have distinct labels or outcomes. Conversely, *vertical federated learning* involves datasets with overlapping feature attributes, but different sample IDs. Here, the data samples may have different input variables but share similar labels. *Federated transfer learning* extends this concept further, incorporating aspects of both horizontal and vertical federated learning. It involves leveraging knowledge or models trained on one dataset to improve learning or performance on another dataset with potentially different characteristics.

Horizontal federated learning divides datasets as it is shown in Fig. 2a. For instance, let's suppose we have three hospitals located in different cities. Hospitals collect data from their patients, without sharing their sensitive data, to train a machine learning model to diagnose a particular disease. In this case, hospitals have datasets that contain the same information, such as patients' age, gender, laboratory test results, and symptoms. The user IDs can overlap, but just to some extent. Consequently, horizontal federated learning facilitates the expansion of the user sample size.

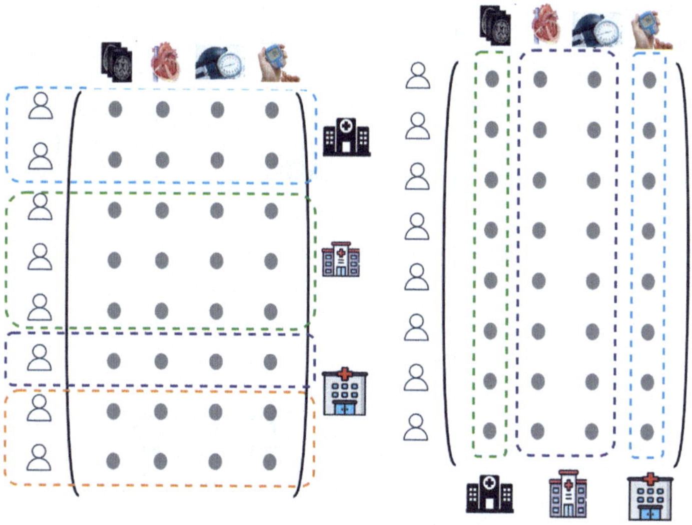

(a) Horizontal federated learning (b) Vertical federated learning

Fig. 2 Comparison between horizontal and vertical federated learning

Vertical federated learning is shown in Fig. 2b. It involves partitioning datasets vertically, where the user features of two datasets can exhibit a certain degree of overlap, while the user IDs overlap considerably. In Fig. 2b hatch hospital has collected data eventually from the same patients, but the data collected by each hospital are inherently different. For example, hospital X has collected a large amount of data on patients related to heart diseases, including demographic data, test results, medical history, and treatment information. At the same time, hospital y is specialized in cancer treatment and has collected detailed data on patients related to specific tumors, chemotherapy treatments, drug responses, and imaging test results. By using vertical federated learning, hospital X and hospital Y can combine their dataset to develop more accurate and generalized models, which exploits also correlation between heterogeneous features, to improve patient care. Essentially, this approach enables different data owners to possess similar samples with varying features. Consequently, vertical federated learning contributes to the expansion of the feature space. In this context, an innovative privacy-preserving tree-boosting system, known as *SecureBoost*, was introduced in [20]. This system enables multiple parties with shared user groups, but distinct feature sets, to collaboratively conduct the training process. Notably, *SecureBoost* achieves a comparable level of accuracy to centralized training approaches while safeguarding data privacy.

Federated transfer learning was introduced in [39], to conduct effective federated learning when both the sample and feature fields of the application have minimal overlaps, This approach establishes one representation using limited common sample sets and shared features. This framework combines federated learning, which constructs a machine learning model, with transfer learning, which leverages reusable

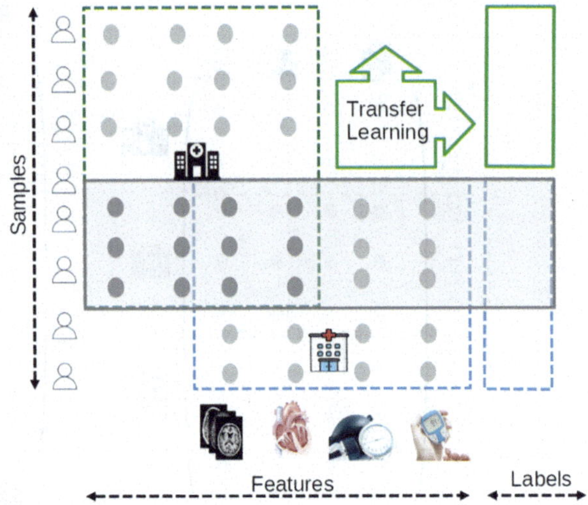

Fig. 3 Federated transfer learning

knowledge across diverse domains. In Fig. 3 two hospitals, specialized in different medical domain, have collected extensive patient data for their respective specialties. Hospital A has gathered data on its patients, but miss feature to develop a machine learning model to predict disease progression and recommend personalized treatment plans. Hospital B has collected data on its patients with cardiovascular diseases. It collected an extend set of features, but share with hospital A a limited subset of samples. Exploiting the intersection of datasets, hospital A leverages the knowledge embedded in Hospital B's model to improve its heart disease prediction model on its own exclusive samples.

5 MAS-Based Federated Learning in Decentralized Environments

Distributed learning and parallel computing frameworks typically operate under the assumption that data are independently and uniformly distributed across different devices, following the principle of independent and identically distributed (i.i.d.) data. However, in federated learning settings, such as those encountered in decentralized environments like edge devices or IoT networks, the data residing on various devices may exhibit imbalances and non-i.i.d. characteristics. These disparities present significant statistical barriers when attempting to construct a unified global model that accurately represents the underlying data distribution across all devices [48]. Moreover, in addition to the challenges stemming from heterogeneous data distribution, the participating devices may also differ significantly in terms of their computational capabilities, storage capacity, and network connectivity. Such variations in hardware infrastructure complicate further the federated learning process, potentially resulting in discrepancies in model performance and convergence across devices. Moreover, the heterogeneous nature of device hardware introduces concerns regarding reliability, as certain devices may be less dependable or more prone to errors compared to others. Consequently, these factors collectively contribute to significant challenges in ensuring fault tolerance and achieving robust model convergence in federated learning environments. Addressing these challenges requires the development of adaptive strategies to manage data heterogeneity effectively, optimize resource allocation, and enhance the resilience of federated learning systems in dynamic and diverse operational contexts [5]. Existing solutions focus on accelerating convergence, minimizing processing latency, or reducing communication overhead individually, often neglecting the interplay between these objectives.

Multi-agent system (MAS) represent a powerful model to solve distributed computation problems, including being able to adapt their operation in open and dynamic environments. Multi-agent coordination has gained significant development in theoretical research in parallel with the increasing attention in the practical applications of such systems in various areas.

We identify here three pillars which leverage the adoption of MAS to develop FL solutions: *Volunteer computing*, *Interaction protocols* and *Reinforcement learning*.

5.1 Volunteer Computing

Volunteer Computing (VC) is a kind of distributed computing that harnesses the aggregated spare computing resources of volunteer devices [33].Volunteer computing has been designed to allow anyone with a computer to donate idle computing resources to run computationally and storage-intensive tasks. In the context of federated learning, volunteers are recruited for knowledge sharing, leveraging the collective intelligence embedded within the global model to derive insights or make predictions relevant to their specific applications or tasks. However, similar to the VC paradigm, volunteers must also provide computational resources for extracting such knowledge from local data. For this reason, functionalities provided by VC middlewares can be exploited to develop the backbone of FL platform, such as the partitioning of volunteer tasks into subtasks so that they can run on volunteer nodes efficiently, the optimal scheduling of tasks and resources to enhance performance and improve the robustness of the overall system, effective communication among a vast number of intermittently available nodes, providing trust and security services, etc. Moreover, VC middlewares support a mechanism to attract and retain volunteers. The *IEEE Std 3652.1-2020* standard provide hints about how these mechanisms can be used to leverage the collective contributions of data owners, coordinators, model users, and auditors, federated learning endeavors to harness the power of decentralized data while maintaining stringent privacy and security requirements. Agents can dynamically enter or leave the system, ensuring flexibility and openness. The proactivity of intelligent software agents can be exploited to delegate the utilization of such mechanism on the user's behalf, according to an high level policy, which defines the goals and constraints to enable the activation and autonomous pursuit of the required plans, to accomplish the final emergent behavior.

5.2 Interaction Protocols

Negotiation models have emerged as suitable candidates for solving the interaction problem due to their decentralized nature, emphasis on mutual selection of actions, and prevalence in real social systems. For this reason, the distributed coordination among a large number of nodes managed by autonomous agents, as well as the coordination between the central FL aggregator and the nodes themselves, can be addressed through the design and engineering of a negotiation model for autonomous agents. This model facilitates the sharing of tasks and resources. In recent years, multi-agent coordination has gained significant development in theoretical research, in parallel with the increasing attention given to the practical applications of such systems in various areas. Therefore, coordination among autonomous nodes, and the interaction between the central FL aggregator and the nodes themselves can be addressed by as the design and the engineering of a negotiation model, complemented with a set of deliberation mechanisms for autonomous agents for sharing tasks and/or resources. In Fig. 4 the interaction between the FL nodes and the selection of the relevant

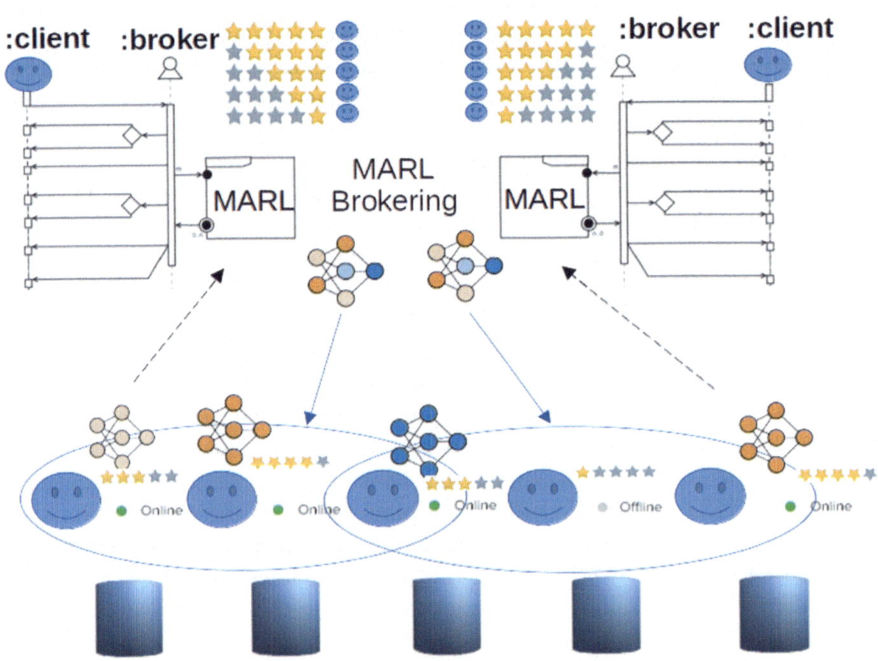

Fig. 4 Multi Agents based FL model

contributors by the central server is modeled through a Negotiation and a Brokering interaction protocols [9, 11]. The FIPA standardization activity has worked to produce public MAS specifications, acting as a key enabler to support interoperability, open service interaction, and to support heterogeneous development [38]. The Agent Communication Language (ACL) and the FIPA (agent interaction) model, often referred to as the FIPA-ACL have been used to define an Agent Interaction Protocol Suite (AIPS). The AIPS contains several distinct semantic protocols for agent communication including: Interaction Process, Communicative Acts, Content Logic, and Content Ontologies.

5.3 Reinforcement Learning

The selection of those devices which promise an effective contribution to the training is a main pillar for accelerating the learning process and improving the learning result. It is very difficult to identify not only the best individuals, but also the design of a policy that is general enough to be valid for a long time and in a dynamic, changing context. Reinforcement learning can be considered a candidate solution to address this problem.

Multi-Agent Reinforcement Learning (MARL) allows the exploit of a distributed paradigm that enables each FL participant to understand themselves and whether their contribution to the collaborative training can be valuable respect to the others. In Fig. 4 MARL is used by the FL aggregator as a client elector that assigns rewards to negotiating agents that handle FL nodes [58]. In particular, the FL aggregator uses the assigned rewards to resolve the brokering issue, aiming at maximizing a global reward. On their own, FL nodes use their reward implementing a selfish behaviour, but in a cooperative setting contributing to maximize the global one.

6 Related Work

We close this chapter with a state-of-the-art review on federated learning, aiming to classify selected publications according to the main objective of each contribution and the topics addressed by the authors, with a special focus on the e-health domain. The results of this study are summarized in Table 1.

Several papers provide a general overview of federated learning. In [46], the authors describe an iterative federated learning approach in which edge nodes are selected to train the current model on their local data. This work also presents various applications of federated learning across different domains and includes a scouting of publicly available datasets. It addresses complexities and issues in implementing FL solutions, such as resource allocation, data imbalance, statistical heterogeneity, costly communication, and privacy concerns. An analysis of the requirements for designing an end-to-end FL infrastructure focused on healthcare use cases is found in [40].

Table 1 Survey studies on federate learning

Paper title	Objective	Topics	Year
Reference [46]	Overview	Model, applications, algorithms, dataset and issues	2021
Reference [1]	Healthcare context	Architectures, applications, algorithms and challenges	2020
Reference [42]	Healthcare context	Structure, devices, applications, datasets and security	2023
Reference [41]	Healthcare context	e-health, taxonomy	2023
Reference [3]	Healthcare application	e-health (3.0, 4.0, 5.0), use case	2023
Reference [44]	Healthcare application	Well-being, architectures, taxonomy, use case	2022
Reference [7]	Healthcare application	Use case, methodologies	2022
Reference [57]	Overview	Model features, challenges and applications	2021
Reference [22]	IoT	Taxonomy, metrics	2021
Reference [2]	IoT and blockchain	Applications, taxonomy	2021
Reference [15]	Framework comparison	Benchmark	2022
Reference [40]	Healthcare context	Framework	2023
Reference [19]	Framework comparison	Strengths, weaknesses	2023
Reference [27]	Framework illustration	Implementation, application and performance	2023
Reference [21]	Framework illustration	Implementation and performance	2021
Reference [29]	Framework illustration	Implementation, application and use case	2023
Reference [35]	Frameworks illustration	Implementations, applications and use cases	2023
Reference [55]	Framework illustration	Implementation, applications and use cases	2023
Reference [50]	Framework illustration	Implementation, applications and use cases	2023
Reference [4]	Framework illustration	Implementation and application	2023
Reference [61]	Framework illustration		2020
Reference [54]	Framework illustration	Implementation and application	2023
Reference [36]	Framework illustration	Implementation and application	2023
Reference [34]	Volunteer computing	Algorithm and application	2022
Reference [30]	Volunteer computing	Algorithm and application	2023
Reference [14]	Healthcare context	Applications, algorithms, dataset	2023
Reference [53]	Healthcare context	Applications, algorithms, dataset	2022
Reference [56]	Healthcare context	Applications, algorithms, dataset	2022
Reference [60]	Healthcare context	Applications, algorithms, dataset	2022
Reference [1]	Software and hardware platforms	Architecture, applications, edge devices	2020
Reference [34]	Platform	Deep learning, volunteer edge devices	2022
Reference [6]	Platform	Architecture, applications	2022
Reference [31]	Platform	Architecture,applications	2023
Reference [25]	Platform	Architecture,applications	2021

FL algorithms represent a key topic in the study. In [1], a study compares various algorithms, including FedAvg, FSVRG, FedProx, and specifically in the medical domain, LoAdaBoost and BrainTorrent. The study examines different algorithm optimization techniques such as mini-batch gradient descent in FL, AdaGrad, RMSProp, and Adam. Similarly, [46] proposes the federated averaging algorithm FedAvg and compares it with other federated learning algorithms. In [14], the FedImpPSO algorithm, an enhanced version of Particle Swarm Optimization (PSO), is recommended and evaluated using the CIFAR-10 and CIFAR-100 datasets, along with a convolutional neural network (CNN). Reference [53] introduces the FedSTCN algorithm, which uses spatio-temporal convolutional networks (STCNs) based on FL to extract temporal and spatial features, improving prediction accuracy of multisensory (IoT) data. The Multi-Layer Stochastic Gradient Descent (MLSGD) algorithm is discussed in [56], aiming to achieve the best global model on e-health data by aggregating partial models learned on peripheral nodes before exchanging them with hospitals. The NET-FLEET algorithm, proposed in [60], enhances the local update scheme in federated learning by incorporating a recursive gradient correction technique to handle heterogeneous datasets. Reference [17] investigates the use of generative AI models, including Generative Adversarial Networks (GANs) and Variational Autoencoders (VAEs), to generate synthetic, realistic, and anonymized patient data, protecting privacy and enabling research and training in the healthcare sector.

Architectures and frameworks for federated learning are another relevant research topic. In the healthcare context, [1] presents the utilization of different types of FL architectures, including horizontal FL, vertical FL, Multi-Participant Multi-class Vertical Federated Learning Framework (MMVFL), Federated Transfer Learning (FTL), and others. This study highlights FL-related challenges such as data distortion, communication, security, privacy, and model accuracy, and analyzes specific use cases including Patient Clustering, Drug Discovery, fMRI Analysis, Brain Tumor Segmentation, Distributed Medical Databases, Medical Entity Recognition, and Medical Imaging. Reference [26] proposes a new collaborative eHealth system called Multilevel Privacy-Preserving Data Sharing (MPPDS), supporting multilevel data sharing while maintaining different levels of privacy protection. ModularFed, described in [4], addresses challenges related to FL implementation complexity by providing a modular and flexible structure to integrate different FL solutions while maintaining extensibility and scalability. FLScalize, detailed in [54], enables researchers to apply their data and models to FL environments, simplifying distribution and management of the FL lifecycle. Reference [36] presents TKAGFL, which uses federated GANs to address data heterogeneity, employs homomorphic encryption to balance data sharing and privacy protection, and combines classical AdaGrad optimization in deep learning with top-K algorithms to compress communication parameters and enhance communication efficiency. Reference [34] proposes an algorithm for Federated Edge Learning (FEEL), combining edge computing and federated learning for volunteer computing on edge devices, and demonstrates good adaptability, dynamism, and error tolerance through various experiments.

The type of participation is another feature distinguishing FL approaches. Reference [30] focuses on the cross-silo federated learning context, where clients are institutions like hospitals, universities, or companies, each with the computational power to train a generative model locally. Reference [29] introduces personalized federated learning based on contribution-aware federated learning (CAreFL), integrating methods for evaluating client contributions and analyzing time series data from mobile and embedded sensing applications [35, 50].

Internet of Things (IoT) is another fundamental research topic for federated learning applications, especially in the medical field. Readers can refer to [19, 22] for surveys focusing on IoT applications. Distributed federated learning (DFL) [21] and decentralized federated learning (PPVD-FL) [27] are investigated in contexts where no centralized server is used. Blockchain technology is proposed in several works for secure, reliable, and optimized IoT data sharing, as argued in [2]. Decentralized frameworks based on intelligent agents that preserve privacy and allow agent participation in both the blockchain network and FL activities are presented in [10, 61]. Reference [42] investigates the security of Internet of Medical Things (IoMT) applications based on FL. Reference [55] introduces Similarity-Quality-based Messenger Distillation (SQMD) for asynchronous heterogeneous collaborative learning on IoT healthcare devices.

A comprehensive work on the progress and evolution of e-health due to the integration of IoT devices, big data, and artificial intelligence into healthcare systems is presented in [3]. This work focuses on technologies like AI for predictive analysis, IoT for real-time patient monitoring, and FL and blockchain to preserve data privacy, creating an intelligent and interconnected healthcare system that promotes patient-centered care. Case studies discussed include predictive modeling for pandemic preparedness, international collaboration on vaccine development, diagnostic imaging for pandemic response, a global infectious disease surveillance system, AI-assisted remote patient monitoring, and privacy-respecting contact tracing.

Finally, existing platforms for developing and executing federated learning algorithms are considered. TensorFlow Federated (TFF[1]) offers a flexible and open framework for users' federated learning needs. PySyft focuses on privacy, managing private data during model training using federated learning within PyTorch [1]. KubeFL [25] streamlines the deployment of FL across multiple clients. Federated AI Technology Enabler (FATE[2]) focuses on security, providing a secure computing architecture based on cryptography. LEAF[3] supports FL, multitasking, and meta-learning with various datasets for experimentation. Reference [34] proposes a software platform for distributed deep learning on volunteer edge devices. The FeatureCloud platform [31] supports multi-institutional FL analytics for developers and end-users. Substra[4] and Flower[5] are recent open-source solutions gaining attention in the research community.

---

[1] https://www.tensorflow.org/federated.

[2] https://fate.fedai.org/federated-network/.

[3] https://leaf.cmu.edu/.

[4] https://github.com/Substra.

[5] https://flower.ai/.

In specific research endeavors, we draw attention to the framework outlined in [47], where researchers advocate for the integration of blockchain and federated learning methodologies to construct a robust architecture tailored for safeguarding privacy in smart healthcare applications. Here, the aim is to bolster security and confidentiality by synergistic strong mode harnessing the capabilities of both technologies. Lastly, an intriguing investigation detailed in [45] proposes a novel algorithm aimed at implementing consensus mechanisms for federated learning within large-scale network environments, with validation conducted in an industrial IoT setting. Major players in the mobile industry are not solely engaged in researching federated learning but are actively implementing this algorithm in various applications. Notably, Google has integrated federated learning into its Gboard mobile keyboard [13], Android messaging platform [18], and certain functionalities within the Pixel smartphone lineup. Similarly, Apple has embraced federated learning, incorporating it into its QuickType mobile keyboard and even the voice classifier for "Hey, Siri" across devices running iOS [37].

FLaMAS is a new federated learning tool for multi-agent systems based on SPADE. FLaMAS brings together the benefits of deep learning and multi-agent systems to create a new distributed learning tool. The system allows the training of deep learning models from anywhere in the world, following the main features of a federated learning algorithm, being distributed (enhanced by the implementation in SPADE agents), and ensuring privacy of the data (as only the model being learned is shared and not the data used for the training).

In their work cited as [24], the authors conduct a thorough examination and comparative assessment of various open-source federated learning frameworks. They meticulously evaluate these frameworks' suitability for integration into IoT systems, considering factors such as analytical capabilities, precision, deployment ease, developmental aspects, and performance metrics. Similarly, another noteworthy review outlined in [23] scrutinizes recent advancements in federated learning with implications for IoT applications. This review introduces a range of evaluative metrics, including sparsification, robustness, quantization, scalability, as well as security and privacy considerations, to gauge the efficacy of the examined frameworks. Furthermore, an additional review referenced as [59] delves into the intricacies of federated learning within IoT platforms, elucidating both the opportunities and challenges inherent in this domain. The paper not only identifies critical hurdles but also proposes innovative strategies to overcome them, shedding light on promising avenues for future exploration.

7 Conclusions

Personal data privacy is a critical concern in today's digital age, with individuals and organizations alike prioritizing the protection of sensitive information. The traditional approach of centralizing data on servers for machine learning model training poses significant challenges in terms of privacy and security. However, federated learning emerges as a groundbreaking solution to address these issues and deliver enhanced performance to users and organizations. At its core, federated learning revolutionizes the conventional model training paradigm by allowing models to be trained directly on local data stored within individual devices. Unlike centralized approaches, where raw data is transferred to a central server for processing, federated learning ensures that data remains decentralized and secure within the confines of users' devices. This decentralized approach alleviates concerns regarding data exposure and privacy breaches, as only model updates or outputs are shared with the central server rather than the raw data itself. By leveraging the collective knowledge gleaned from locally stored data across a distributed network of devices, federated learning enables the development of robust and accurate machine learning models without compromising individual privacy. This decentralized training methodology empowers users and organizations to harness the wealth of data at their disposal while adhering to strict privacy regulations and maintaining data sovereignty. Moreover, federated learning offers a host of benefits beyond privacy preservation. By training models directly on local data, federated learning minimizes the need for extensive data transfers over networks, resulting in reduced bandwidth consumption and faster model training times. Additionally, the localized nature of federated learning enhances scalability, allowing for the seamless integration of new devices or data sources into the federated learning ecosystem. Furthermore, the collaborative nature of federated learning fosters a spirit of cooperation among participants, enabling the collective improvement of machine learning models through shared insights and experiences. This collaborative approach enhances model performance and fosters innovation and knowledge sharing within the federated learning community. In summary, federated learning represents a paradigm shift in machine learning model training, offering a privacy-preserving, efficient, and collaborative approach to harnessing the power of decentralized data. By embracing federated learning, users and organizations can unlock the full potential of their data assets while safeguarding privacy and driving advancements in machine learning and artificial intelligence. Technical and technological challenges arise, but related work demonstrate how the research effort can help to leverage the adoption of this promising paradigm. Its strengths may lead to a turning point in driving user acceptance, which is necessary to push learning capabilities beyond current possibilities.

Acknowledgements This work has been supported by the REDRAW research project (P2022MWE3S - Prin 2022 PNRR, DR n. 1409 of 14-09-2022) funded by the Italian Ministry of Research and by the European Union.

References

1. Aledhari M, Razzak R, Parizi RM, Saeed F (2020) Federated learning: a survey on enabling technologies, protocols, and applications. IEEE Access 8:140699–140725. https://doi.org/10.1109/ACCESS.2020.3013541
2. Ali M, Karimipour H, Tariq M (2021) Integration of blockchain and federated learning for internet of things: Recent advances and future challenges. Comput

Sec 108:102355. https://doi.org/10.1016/j.cose.2021.102355. https://www.sciencedirect.com/science/article/pii/S0167404821001796
3. Alsamhi S, Hawbani A, Shvetsov A, Kumar S (2023) Advancing pandemic preparedness in healthcare 5.0: A survey of federated learning applications. Adv Hum-Comput Interact 2023:1–19. https://doi.org/10.1155/2023/9992393
4. Arafeh M, Otrok H, Ould-Slimane H, Mourad A, Talhi C, Damiani E (2023) Modularfed: Leveraging modularity in federated learning frameworks. Internet of Things 22:100694. https://doi.org/10.1016/j.iot.2023.100694. https://www.sciencedirect.com/science/article/pii/S2542660523000173
5. Bonawitz K, Eichner H, Grieskamp W, Huba D, Ingerman A, Ivanov V, Kiddon C, Konečný J, Mazzocchi S, McMahan HB, Overveldt TV, Petrou D, Ramage D, Roselander J (2019) Towards federated learning at scale: system design
6. Burlachenko K, Horváth S, Richtárik P (2022) Fl_pytorch: optimization research simulator for federated learning. CoRR **abs/2202.03099**. https://arxiv.org/abs/2202.03099
7. Camajori Tedeschini B, Savazzi S, Stoklasa R, Barbieri L, Stathopoulos I, Nicoli M, Serio L (2022) Decentralized federated learning for healthcare networks: a case study on tumor segmentation. IEEE Access 10:8693–8708. https://doi.org/10.1109/ACCESS.2022.3141913
8. Chandrasekaran V, Banerjee S, Perino D, Kourtellis N (2022) Hierarchical federated learning with privacy. CoRR **abs/2206.05209** . https://doi.org/10.48550/ARXIV.2206.05209
9. Charlton P, Cattoni R, Potrich A, Mamdani E (2000) Evaluating the fipa standards and their role in achieving cooperation in multi-agent systems. In: Proceedings of the 33rd annual hawaii international conference on system sciences, pp 10 pp. vol 2–. https://doi.org/10.1109/HICSS.2000.926996
10. Cheng Z, Pan Y, Liu Y, Wang B, Deng X, Zhu C (2022) Vflchain: blockchain-enabled vertical federated learning for edge network data sharing. In: 2022 IEEE international conference on unmanned systems (ICUS), pp 606–611. https://doi.org/10.1109/ICUS55513.2022.9987097
11. Cuomo A, Di Modica G, Distefano S, Puliafito A, Rak M, Tomarchio O, Venticinque S, Villano U (2013) An sla-based broker for cloud infrastructures. J Grid Comput 11(1):1–25. https://doi.org/10.1007/s10723-012-9241-4. Cited by: 68
12. Di Fatta G, Fortino G (2007) A customizable multi-agent system for distributed data mining. In: Proceedings of the 2007 ACM symposium on applied computing, SAC '07, pp 42–47. Association for Computing Machinery, New York, NY, USA. https://doi.org/10.1145/1244002.1244012
13. Hard A, Rao K, Mathews R, Ramaswamy S, Beaufays F, Augenstein S, Eichner H, Kiddon C, Ramage D (2019) Federated learning for mobile keyboard prediction
14. Houssein EH, Sayed A (2023) Boosted federated learning based on improved particle swarm optimization for healthcare IoT devices. Comput Biol Med 163:107195. https://doi.org/10.1016/j.compbiomed.2023.107195. https://www.sciencedirect.com/science/article/pii/S0010482523006601
15. Hu S, Li Y, Liu X, Li Q, Wu Z, He B (2022) The oarf benchmark suite: Characterization and implications for federated learning systems. ACM Trans Intell Syst Technol 13(4). https://doi.org/10.1145/3510540
16. Inc. CS (2020) The role of technology in powering an inclusive future. https://www.cisco.com/c/dam/m/en_us/about/cxo-agenda/inclusive-future/the-role-of-technology-in-powering-an-inclusive-future.pdf
17. Jadon A, Kumar S (2023) Leveraging generative AI models for synthetic data generation in healthcare: Balancing research and privacy. In: 2023 international conference on smart applications, communications and networking (SmartNets). IEEE (2023). https://doi.org/10.1109/smartnets58706.2023.10215825
18. Kairouz, P., McMahan, H.B., Avent, B., Bellet, A., Bennis, M., Bhagoji, A.N., Bonawitz, K., Charles, Z., Cormode, G., Cummings, R., D'Oliveira, R.G.L., Eichner, H., Rouayheb, S.E., Evans, D., Gardner, J., Garrett, Z., Gascón, A., Ghazi, B., Gibbons, P.B., Gruteser, M., Harchaoui, Z., He, C., He, L., Huo, Z., Hutchinson, B., Hsu, J., Jaggi, M., Javidi, T., Joshi, G., Khodak, M., Konečný, J., Korolova, A., Koushanfar, F., Koyejo, S., Lepoint, T., Liu, Y.,

Mittal, P., Mohri, M., Nock, R., Özgür, A., Pagh, R., Raykova, M., Qi, H., Ramage, D., Raskar, R., Song, D., Song, W., Stich, S.U., Sun, Z., Suresh, A.T., Tramèr, F., Vepakomma, P., Wang, J., Xiong, L., Xu, Z., Yang, Q., Yu, F.X., Yu, H., Zhao, S.: Advances and open problems in federated learning (2021)
19. Karimireddy SP, Veeraragavan NR, Elvatun S, Nygård JF (2023) Federated learning showdown: the comparative analysis of federated learning frameworks. In: 2023 eighth international conference on fog and mobile edge computing (FMEC), pp 224–231. https://doi.org/10.1109/FMEC59375.2023.10305961
20. Kewei C, Tao F, Yilun J, Yang L, Tianjian C, Papadopoulos D, Qiang Y (2021) Secureboost: a lossless federated learning framework
21. Khan LU, Saad W, Han Z, Hong CS (2021) Dispersed federated learning: vision, taxonomy, and future directions. IEEE Wirel Commun 28(5):192–198. https://doi.org/10.1109/MWC.011.2100003
22. Khan LU, Saad W, Han Z, Hossain E, Hong CS (2021) Federated learning for internet of things: recent advances, taxonomy, and open challenges. IEEE Commun Surv Tutor 23(3):1759–1799. https://doi.org/10.1109/COMST.2021.3090430
23. Khan LU, Saad W, Han Z, Hossain E, Hong CS (2021) Federated learning for internet of things: recent advances, taxonomy, and open challenges
24. Kholod I, Yanaki E, Fomichev D, Shalugin E, Novikova E, Filippov E, Nordlund M (2021) Open-source federated learning frameworks for IoT: a comparative review and analysis. Sensors 21(1). https://doi.org/10.3390/s21010167
25. Kim J, Kim D, Lee J (2021) Design and implementation of kubernetes enabled federated learning platform. In: 2021 international conference on information and communication technology convergence (ICTC), pp 410–412. https://doi.org/10.1109/ICTC52510.2021.9620986
26. Kim JW, Edemacu K, Jang B (2019) Mppds: multilevel privacy-preserving data sharing in a collaborative ehealth system. IEEE Access 7:109910–109923. https://doi.org/10.1109/ACCESS.2019.2933542
27. Kumar SS, Joshith TS, Lokesh DD, Jahnavi D, Mahato GK, Chakraborty SK (2023) Privacy-preserving and verifiable decentralized federated learning. In: 2023 5th international conference on energy, power and environment: towards flexible green energy technologies (ICEPE), pp 1–6 (2023). https://doi.org/10.1109/ICEPE57949.2023.10201599
28. Li T, Sahu AK, Talwalkar A, Smith V (2020) Federated learning: Challenges, methods, and future directions. IEEE Signal Process Mag 37:50–60. http://dx.doi.org/10.1109/MSP.2020.2975749
29. Liu Z, Chen Y, Zhao Y, Yu H, Liu Y, Bao R, Jiang J, Nie Z, Xu Q, Yang Q (2023) Carefl: enhancing smart healthcare with contribution-aware federated learning. AI Mag 44(1):4–15. https://doi.org/10.1002/aaai.12082
30. Lomurno E, Archetti A, Cazzella L, Samele S, Di Perna L, Matteucci M (2023) Sgde: Secure generative data exchange for cross-silo federated learning. In: Proceedings of the 2022 5th international conference on artificial intelligence and pattern recognition, AIPR '22, pp 205–214. Association for Computing Machinery, New York, NY, USA. https://doi.org/10.1145/3573942.3573974
31. Matschinske J, Späth J, Bakhtiari M, Probul N, Majdabadi M, Nasirigerdeh R, Torkzadehmahani R, Hartebrodt A, Orban BA, Fejér SJ, Zolotareva O, Das S, Baumbach L, Pauling J, TomaševiĆ O, Bihari B, Bloice M, Donner N, Fdhila W, Baumbach J (2023) The feature cloud platform for federated learning in biomedicine: unified approach. J Med Internet Res 25:e42621. https://doi.org/10.2196/42621
32. McHugh KA, Yammarino FJ, Dionne SD, Serban A, Sayama H, Chatterjee S (2016) Collective decision making, leadership, and collective intelligence: tests with agent-based simulations and a field study. The Leadership Quart 27(2):218–241. https://doi.org/10.1016/j.leaqua.2016.01.001. https://www.sciencedirect.com/science/article/pii/S1048984316000023. Special Issue: Collective and Network Approaches to Leadership
33. Mengistu TM, Che D (2019) Survey and taxonomy of volunteer computing. ACM Comput Surv **52**(3). https://doi.org/10.1145/3320073

34. Ángel Morell J, Alba E (2022) Dynamic and adaptive fault-tolerant asynchronous federated learning using volunteer edge devices. Future Generat Comput Syst 133:53–67 (2022). https://doi.org/10.1016/j.future.2022.02.024. https://www.sciencedirect.com/science/article/pii/S0167739X22000735
35. Park J, Lee K, Lee S, Zhang M, Ko J (2023) Attfl: a personalized federated learning framework for time-series mobile and embedded sensor data processing. Proc ACM Interact Mob Wearable Ubiquitous Technol 7(3). https://doi.org/10.1145/3610917
36. Pei J, Yu Z, Li J, Jan MA, Lakshmanna K (2023) Tkagfl: a federated communication framework under data heterogeneity. IEEE Trans Netw Sci Eng 10(5):2651–2661. https://doi.org/10.1109/TNSE.2022.3177748
37. Peterson D, Kanani P, Marathe VJ (2019) Private federated learning with domain adaptation
38. Poslad S, Charlton P (2001) Standardizing agent interoperability: the FIPA approach. Springer, Berlin, Heidelberg, pp 98–117. https://doi.org/10.1007/3-540-47745-4_5
39. Qiang Y, Yang L, Tianjian C, Yongxin T (2019) Federated machine learning: concept and applications
40. Rachakonda S, Moorthy S, Jain A, Bukharev A, Bucur A, Manni F, Quiterio TM, Joosten L, Mendez NI (2023) Privacy enhancing and scalable federated learning to accelerate AI implementation in cross-silo and iomt environments. IEEE J Biomed Health Inf 27(2):744–755. https://doi.org/10.1109/JBHI.2022.3185418
41. Rahman A, Hossain M, Muhammad G, Kundu D, Debnath T, Rahman M, Khan M, Tiwari P, Band S (2023) Federated learning-based AI approaches in smart healthcare: concepts, taxonomies, challenges and open issues. Cluster Comput 26(4):2271–2311. https://doi.org/10.1007/s10586-022-03658-4. Publisher Copyright: 2022, The Author(s), under exclusive licence to Springer Science+Business Media, LLC, part of Springer Nature
42. Rani S, Kataria A, Kumar S, Tiwari P (2023) Federated learning for secure iomt-applications in smart healthcare systems: a comprehensive review. Knowl-Based Syst 274:110658. https://doi.org/10.1016/j.knosys.2023.110658. https://www.sciencedirect.com/science/article/pii/S0950705123004082
43. Rincon J, Julian V, Carrascosa C (2022) Flamas: federated learning based on a spade mas. Appl Sci 12(7). https://doi.org/10.3390/app12073701
44. Saraswat D, Bhattacharya P, Verma A, Prasad VK, Tanwar S, Sharma G, Bokoro PN, Sharma R (2022) Explainable AI for healthcare 5.0: opportunities and challenges. IEEE Access 10:84486–84517. https://doi.org/10.1109/ACCESS.2022.3197671
45. Savazzi S, Nicoli M, Rampa V (2020) Federated learning with cooperating devices: a consensus approach for massive IoT networks. IEEE Internet of Things J 7(5):4641–4654. http://dx.doi.org/10.1109/JIOT.2020.2964162
46. Shaheen M, Farooq MS, Umer T, Kim BS (2022) Applications of federated learning; taxonomy, challenges, and research trends. Electronics 11(4). https://doi.org/10.3390/electronics11040670
47. Singh S, Rathore S, Alfarraj O, Tolba A, Yoon B (2021) A framework for privacy-preservation of IoT healthcare data using federated learning and blockchain technology. Future Generat Comput Syst 129 (2021). https://doi.org/10.1016/j.future.2021.11.028
48. Smith, V., Chiang, C.K., Sanjabi, M., Talwalkar, A.S.: Federated multi-task learning. In: I. Guyon, U.V. Luxburg, S. Bengio, H. Wallach, R. Fergus, S. Vishwanathan, R. Garnett (eds.) Advances in Neural Information Processing Systems, vol. 30. Curran Associates, Inc. (2017). URLhttps://proceedings.neurips.cc/paper_files/paper/2017/file/6211080fa89981f66b1a0c9d55c61d0f-Paper.pdf
49. Suran S, Pattanaik V, Draheim D (2020) Frameworks for collective intelligence: a systematic literature review. ACM Comput Surv 53(1). https://doi.org/10.1145/3368986
50. Tashakori A, Zhang W, Jane Wang Z, Servati P (2023) Semipfl: personalized semi-supervised federated learning framework for edge intelligence. IEEE Internet of Things J 10(10):9161–9176. https://doi.org/10.1109/JIOT.2022.3233599
51. Vepakomma P, Gupta O, Swedish T, Raskar R (2018) Split learning for health: Distributed deep learning without sharing raw patient data (2018)

52. Wainakh A, Guinea AS, Grube T, Mühlhäuser M (2020) Enhancing privacy via hierarchical federated learning. In: 2020 IEEE European symposium on security and privacy workshops (EuroS&PW) pp 344–347. https://api.semanticscholar.org/CorpusID:216080863
53. Xing Y, An H, Zhang Y (2022) Federated learning-based space-time data prediction on edge devices. In: 2022 7th international conference on intelligent computing and signal processing (ICSP), pp 737–741. https://doi.org/10.1109/ICSP54964.2022.9778747
54. Yang S, Moon J, Kim J, Lee K, Lee K (2023) Flscalize: federated learning lifecycle management platform. IEEE Access 11:47212–47222. https://doi.org/10.1109/ACCESS.2023.3275439
55. Ye G, Chen T, Li Y, Cui L, Nguyen QV, Yin H (2023) Heterogeneous collaborative learning for personalized healthcare analytics via messenger distillation
56. Yu C, Shen S, Wang S, Zhang K, Zhao H (2022) Efficient multi-layer stochastic gradient descent algorithm for federated learning in e-health. In: ICC 2022 - IEEE international conference on communications, pp 1263–1268. https://doi.org/10.1109/ICC45855.2022.9839100
57. Zhang C, Xie Y, Bai H, Yu B, Li W, Gao Y (2021) A survey on federated learning. Knowl-Based Syst 216:106775. https://doi.org/10.1016/j.knosys.2021.106775. https://www.sciencedirect.com/science/article/pii/S0950705121000381
58. Zhang SQ, Lin J, Zhang Q (2022) A multi-agent reinforcement learning approach for efficient client selection in federated learning
59. Zhang T, Gao L, He C, Zhang M, Krishnamachari B, Avestimehr AS (2022) Federated learning for internet of things: applications, challenges, and opportunities
60. Zhang X, Fang M, Liu Z, Yang H, Liu J, Zhu Z (2022) Net-fleet: achieving linear convergence speedup for fully decentralized federated learning with heterogeneous data
61. Zhang Z, Yang T, Liu Y (2020) Sablockfl: a blockchain-based smart agent system architecture and its application in federated learning. Int J Crowd Sci 4(2):133–147. https://doi.org/10.1108/IJCS-12-2019-0037

Applications of Solid Platform and Federated Learning for Decentralized Health Data Management

Franco Cirillo, Marco De Santis, and Christian Esposito

Abstract The integration of artificial intelligence (AI) into medical cyber-physical systems has led to novel approaches for managing healthcare data. However, centralizing healthcare data poses risks to privacy and security. Decentralized approaches offer a promising solution to balance the need for information sharing with individual privacy protection. This chapter explores the Solid platform for decentralized healthcare data management, such as electronic health records (EHR) and Federated Learning (FL) that emerges as a technique to extract knowledge from distributed data while preserving user privacy. Despite the benefits, challenges remain in adopting decentralized approaches, including the need for standards, patient and healthcare professional education, and ethical considerations. Moreover, patients play a crucial role in decentralized healthcare data management, necessitating informed involvement in data control decisions. Nonetheless, FL holds promise for personalized medicine, advancing healthcare delivery, enhancing patient outcomes, accelerate medical research and addressing regulatory compliance challenges in the medical field.

1 Introduction

Artificial intelligence (AI) is revolutionizing the way we manage health data [2]. Its ability to analyze large volumes of data offers new possibilities for improving the diagnosis, treatment and prevention of disease. However, the centralization of health data in large databases poses significant risks to the privacy and security of individuals [3].

In this context, decentralized approaches to health AI [1] offer a promising solution to balance the need for information sharing with protecting the privacy of individuals. These approaches rely on distributing health data across multiple devices or servers, avoiding the creation of a single point of vulnerability.

F. Cirillo · M. De Santis (✉) · C. Esposito
University of Salerno, Fisciano, SA, Italy
e-mail: mdesantis@unisa.it

M. Ficco and G. D'Angelo (eds.), *Artificial Intelligence Techniques for Analysing Sensitive Data in Medical Cyber-Physical Systems*, Engineering Cyber-Physical Systems and Critical Infrastructures 12, https://doi.org/10.1007/978-3-031-70775-9_6

In this chapter, we will explore the benefits and challenges of decentralized approaches to health AI. In particular, we will focus on two key technologies:

- Decentralization of Health Data: This concept is pivotal in the realm of medical information management, aiming to shift control and responsibility of data from individual centralized entities to a more distributed structure. By doing so, this approach champions the security and privacy of patients by ensuring their data remains shielded from potential cyber-attacks or misuse by third parties. Moreover, it fosters collaboration among diverse entities within the healthcare sector [4], enabling secure data sharing and analysis for research purposes, treatment development, and clinical practice enhancement. With decentralization, patients also gain greater autonomy over their data, empowering them to grant or revoke access to information in a more transparent and direct manner.
- Decentralization of AI: This aspect delves into the redistribution of AI processes and functionalities from centralized systems to a more distributed architecture. By dispersing AI capabilities across various nodes or devices, this approach enhances scalability, resilience, and efficiency while mitigating single points of failure and privacy concerns associated with centralized models [5]. Decentralized AI empowers edge devices, such as smartphones, wearables, and IoT sensors, to perform AI-related tasks locally, thereby reducing latency, conserving bandwidth, and preserving data privacy. These devices, often employed for patient monitoring, play a crucial role in enabling real-time analysis and feedback, facilitating continuous and personalized healthcare delivery. By leveraging AI algorithms directly on these devices, healthcare providers can monitor vital signs, detect anomalies, and deliver timely interventions without relying on centralized servers or internet connectivity. This not only enhances patient autonomy and convenience but also reduces the burden on healthcare infrastructure and improves the overall quality of care. Furthermore, decentralized AI fosters collaborative learning and decision-making, enabling disparate entities to contribute to AI model training and refinement without compromising data security or integrity.

In this chapter, we will examine both of these approaches in detail, highlighting their potential and challenges, as well as their impact on clinical practice and medical research. We will outline the implications, advantages, and obstacles in the context of healthcare AI. Through the exploration of real-world applications, case studies, and emerging trends, we will seek to provide a comprehensive understanding of the transformational potential and challenges associated with decentralized approaches to healthcare AI.

2 Solid for Decentralized Health Data Management

This section introduces Solid [7], a decentralized platform, highlighting its features such as secure data storage, controlled access, and data sharing capabilities. The paper discusses the advantages of Solid for the management of health data, including improved patient control, enhanced security, and interoperability. We explore the benefits of Solid in this area, including:

- Improved patient control: Patients take complete control of their health data, being able to choose how, when and with whom they share it.
- Advanced Security: Health data is stored securely and decentralized, ensuring confidentiality and integrity of information.
- Interoperability: Health data is structured in an interoperable format, facilitating access and use by different applications and services.

In the digital age, personal data management has become an increasingly complex issue. On the one hand, we desire the convenience of storing our data online so that we can easily access it from any device. On the other, we worry about the privacy and security of our data, and the control that large tech companies have over it.

Decentralization and the semantic web represent two key approaches to addressing these challenges:

1. Decentralization: User data are not stored on one central server, but distributed across a network of servers. This makes the platform more resistant to cyber attacks and gives users more control over their privacy.
2. Semantic Web: User data is stored in a structured and interoperable format, which means it can be easily read and used by different applications and services.

Solid is an innovative decentralized platform that revolutionizes the way users interact with their data online. Based on the principles of connected data and decentralization, Solid gives users unprecedented control over their digital experience [7]. The key concepts of Solid are the Solid Personal Online Dataspace (POD), Solid Server, WebID, Solid Provider, and Access Control Lists:

- Solid Personal Online Dataspace (POD): Each user has a personal online storage space, known as a POD, where data is securely stored in a structured manner. RDF is utilized to ensure semantic richness and interoperability.
- Solid Server: Instances of web servers compliant with Solid specifications manage user accounts and POD instances, facilitating data management operations.
- WebID: A unique identifier associated with individuals or entities, managed by Solid servers, which enables authentication and access control on the platform.
- Solid Provider: Organizations hosting public instances of Solid servers, offering users the infrastructure to create and host their PODs.
- Access-Control List (ACL): Utilizing Web Access Control (WAC), users define access permissions for their data, specifying who can access it and the level of authorization.

With these key concepts, Solid offers users a secure and controlled platform to store, share, and manage their data online. The combination of decentralization and semantic web makes Solid an innovative solution to address the challenges of managing personal data in the digital age.

2.1 Secure Data Storage

One of the main features of Solid is the ability for users to store their data in a secure and decentralized manner. This is made possible by the Solid Personal Online Dataspace (POD) concept, which serves as a personal online storage space for each user [6]. The data within the POD are:

1. Encrypted: At rest and in transit, using industry standard algorithms such as AES.
2. Controlled: Only the user has access to their data, with the ability to granularly define access permissions.
3. Integer: The use of digital signatures ensures the integrity and authenticity of the data.

Here data are stored using advanced security standards and can be expressed in RDF (Resource Description Framework) or with generated RDF metadata. This ensures that the data is structured and semantic, facilitating understanding and interoperability.

2.2 Data Access Control and Selective Data Sharing

Solid puts total control of access to their data in the hands of users. Using Web Access Control (WAC), users can define who has permission to access their data and with what level of authorization. This is done through the use of access control lists (ACLs), which allow users to specify access permissions for individual objects in their PODs. In this way, users can ensure the privacy and security of their online data.

Therefore, Solid offers users the ability to selectively share their data with whomever they wish. Using the granular access controls provided by WAC, users can define exactly who can access their data and how. Through Web Access Control (WAC), users can:

1. Define permissions: Specify who can access their data and with what level of authorization.
2. Selectively share: Share their data with individuals, groups or make it completely public.
3. Revoke access: Remove access permission at any time.

They can grant access to individual users, groups of users or even make the data completely public. In this way, users can maintain control over their privacy while sharing information with others.

2.3 Examples of Solid Applications in Healthcare

This subsection explores various applications of Solid in the medical field, including decentralized electronic health records, medical research based on patient data, and the development of new drugs and therapies.

Solid's potential extends beyond individual data ownership, offering exciting possibilities for the healthcare industry. Here are some key applications:

1. Decentralized Electronic Health Records (EHRs): Solid empowers patients to create personal health vaults within their PODs, securely storing their entire medical history, including diagnoses, medications, allergies, and lab results [8–10]. Data ownership and control reside with patients, granting them the ability to decide who can access their records and for what purpose. Authorized healthcare providers (doctors, nurses, specialists) can gain secure access to a patient's consented and shared data, enabling seamless care coordination and informed decision making. Interoperability with existing healthcare systems can be facilitated through standardized data formats, allowing seamless data exchange between different institutions.
   As depicted in Fig. 1, the scenario illustrates how Solid enables patients to securely manage their medical data within personal PODs while granting authorized healthcare providers access to consented data, facilitating seamless care coordination and interoperability with existing healthcare systems.
2. Medical Research Based on Patient Data: Solid fosters secure and anonymized data collection for medical research. Patients can choose to contribute their data to specific research projects while maintaining complete control over its use and access. Researchers can access relevant and deidentified datasets through approved channels, streamlining research processes and expediting discoveries. Solid's granular access control ensures patient privacy while enabling researchers to gather valuable insights for epidemiological studies, clinical trials, and personalized medicine initiatives [11].
3. Development of New Drugs and Therapies: Solid facilitates secure collaboration between researchers, pharmaceutical companies, and research institutions. Secure data sharing allows for joint analysis of patient data, leading to a deeper understanding of diseases and their mechanisms. Streamlined data access expedites drug discovery and development processes, potentially leading to the creation of more effective and personalized treatments for patients [12]. This revised version provides additional details to showcase the specific functionalities and benefits of Solid in each healthcare application. It also emphasizes patient control, data security, and interoperability aspects that are crucial for successful adoption in the healthcare landscape.

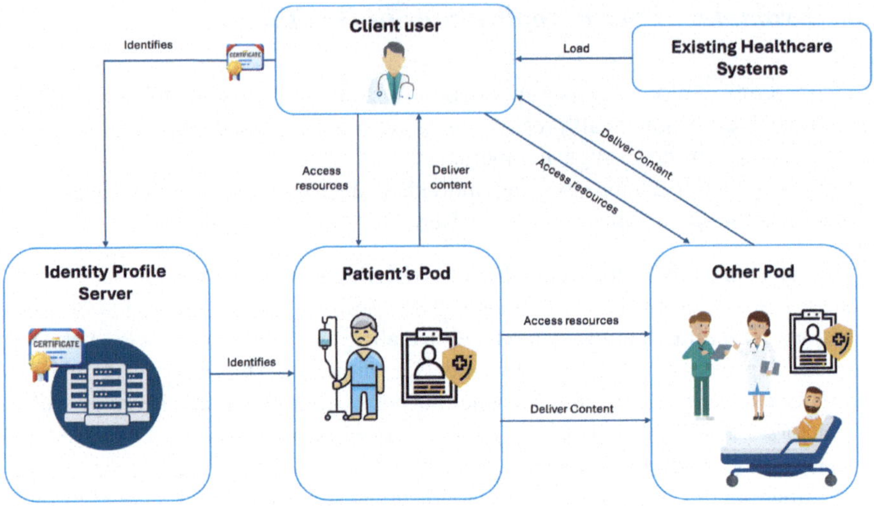

Fig. 1 Decentralized Electronic Health Records (EHR) scenario using solid

3 Federated Learning in the Medical Context

Federated Learning (FL) represents a cutting-edge methodology in machine learning that has shown particular promise in the medical field. This innovative technique enables the training of machine learning models on distributed data on local devices, preserving the privacy of user data while enabling the extraction of meaningful knowledge. In the medical context, where the protection of sensitive patient data is paramount, Federated Learning offers a secure and efficient approach to distributed data analysis [13].

In the medical industry, the use of ML models on Electronic Health Records (EHR) data is becoming increasingly common to improve healthcare decision-making. However, the sensitivity and confidentiality of medical data make it difficult to share such information directly between different entities. FL addresses this challenge by enabling the training of ML models on distributed data on local devices, avoiding the need to centralize data in a single server.

This section discusses FL as a machine learning technique for extracting knowledge from distributed data while maintaining user privacy. Then, it explores applications of FL in healthcare, such as early diagnosis models, personalized treatments, and patient health monitoring.

3.1 Introduction

FL is an advanced machine learning technique designed to train predictive models on distributed data on local devices, without the need to centralise the data in a single server [14]. Instead, the learning model is distributed to remote devices or servers, where each device trains the model using available local data. Subsequently, only minor updates of the model weights are sent to a central server or aggregator, which combines these updates to further improve the overall model. FL comes in different configurations, including:

- Centralised Federated Learning: In this setup, a central server coordinates all phases of the learning algorithm, including the coordination of the participating nodes during the learning process. The server is responsible for the initial selection of nodes and the aggregation of model updates. However, centralization may cause scalability problems.
- Decentralised Federated Learning: In this context, nodes are able to coordinate with each other to obtain the global model without central orchestration. This configuration reduces the risk of single failures, but the network topology may affect overall performance.
- Heterogeneous Federated Learning: With the increasing involvement of heterogeneous clients, such as IoT devices and mobile phones, the need to adapt FL to local models with different architectures and varying computing capabilities has emerged [15].

The FL process has the following key features:

- Iterative learning: The learning process is iterative and divided into federated learning rounds, during which the global model is updated based on the local data of the participating nodes.
- Model aggregation: After the local nodes have trained the model on their own data, updates are sent to the central server or the aggregator for aggregation. This process allows local knowledge to be combined into one global model.
- Asynchronous learning: Some variants of federated learning allow asynchronous model updates, reducing the delay and improving the overall efficiency of the training process.

FL works through a series of well-defined stages:

1. Initialization: A machine learning model is initially deployed on the devices of users participating in the federated learning process.
2. Local Learning: Devices use the initial model to analyze available local data. Model training takes place on local data without the need to transfer the data to a central server.
3. Model update: After local training, devices send only minor updates of model weights to the central server or aggregator.

4. Model aggregation: The central server or aggregator combines model weight updates from participating devices to improve the overall model. This aggregation process takes into account differences in local data and updates the central model to reflect these variations.

These features make FL a powerful methodology for analyzing distributed data in the medical industry, allowing the privacy of patient data to be maintained and meaningful knowledge to be extracted without compromising data security.

3.2 Applications of FL in Medicine

FL) offers a wide range of applications in the medical field, revolutionizing the way artificial intelligence models are developed and utilized to enhance diagnosis, treatment, and patient monitoring [16]. Below are some of the key applications of FL in the medical context:

1. Early Diagnosis and Disease Prediction. FL can be used to develop machine learning models capable of identifying diseases early by analyzing distributed data from Electronic Health Records (EHR) of patients [17, 18]. This approach enables the detection of early disease signals and initiation of timely preventive interventions, significantly improving chances of recovery and reducing the burden on healthcare systems.
2. Personalized Treatments. With the ability to adapt to the specific data of each patient while maintaining data privacy, FL enables treatment personalization based on individual needs, as depicted in Fig. 2 in which local weights updates of different hospital improve a model to enhance a personalized treatment. ML models trained through FL can analyze a wide range of factors, including medical history, responses to previous treatments, and genetic characteristics, to develop optimized treatment plans for each patient [19].
3. Continuous Health Monitoring. By using data from wearable devices and other connected medical sensors, FL can be employed to continuously monitor the health status of patients. ML models can analyze the collected data in real-time, detect anomalies and significant trends, and promptly alert healthcare providers or patients themselves in case of emergencies or changes in clinical status.
4. Clinical Research and Drug Development. FL can also revolutionize the clinical research process and drug development by enabling secure and collaborative data sharing among institutions and organizations. This approach can accelerate the identification of new therapeutic targets, improve patient selection for clinical trials, and optimize the design of clinical trials, leading to more effective and personalized therapies.
5. Disease Prevention and Health Promotion. Finally, FL can be used to develop disease prediction models and identify risk factors associated with certain medical conditions. These models can help identify patients at risk early and adopt targeted

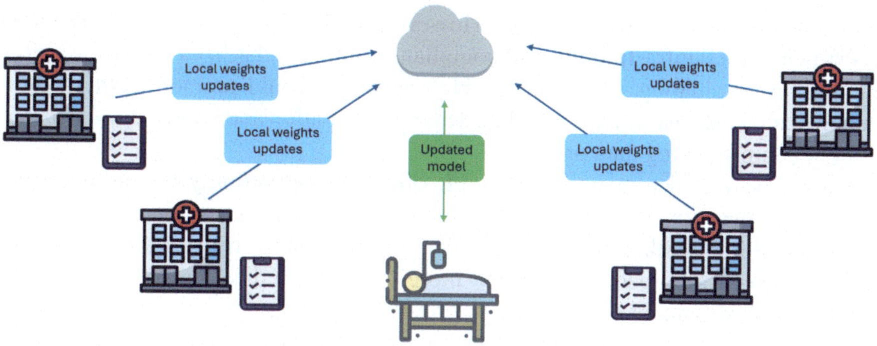

Fig. 2 Application of FL to train models on different hospitals data to improve personalized treatment

preventive interventions to promote health and prevent the development of chronic diseases.

Having discussed the potential applications of FL in the medical field, it is also essential to evaluate all the implications associated with this usage concerning all stakeholders involved with these solutions.

3.3 *Implications for Stakeholders in the Healthcare Sector*

Federated Learning has diverse implications for various stakeholders involved in the treatment cycle, including clinicians, patients, hospitals, healthcare providers, and medical device manufacturers.

For patients, FL can ensure greater data privacy and security, allowing them to be assured that their data remains within their own institution and that access to the data can be revoked. Additionally, FL could lower the barriers to becoming data donors, encouraging greater participation. The ability to actively participate in medical research while retaining control over their data could increase patient empowerment and foster greater trust in the healthcare system.

For clinicians, FL offers better diagnostic tools and decision support, allowing them to access artificial intelligence models trained on a wide collection of real data from various institutions. This could lead to more accurate and timely diagnoses [17, 18], improving patient care. Additionally, FL could support personalized medicine, enabling clinicians to tailor treatments based on specific patient data and best practices emerging from a global data analysis.

For hospitals and healthcare providers, FL allows them to maintain full control over patient data, reducing the risk of abuse by third parties. Moreover, machine learning-based predictive approaches can be extremely beneficial to predict patient's length of stay (LOS), which is a crucial determinant for hospitals to maintain resource efficiency and quality treatment [21]. However, the implementation of FL requires investments in IT infrastructure and compliance with standardized data formats for training and evaluating AI models. Additionally, FL could promote collaboration among healthcare institutions, enabling secure information and knowledge exchange without compromising data privacy.

For medical device manufacturers, FL offers the opportunity to continue validating and improving their AI-based systems without sharing specific patient information, enabling further continuous validation or enhancement of their AI-based systems. This could lead to safer and more efficient medical devices, improving overall healthcare quality.

These implications highlight the transformative potential of FL in revolutionizing healthcare delivery while addressing concerns regarding data privacy, security, and collaboration among stakeholders. As FL continues to evolve and gain traction in the medical field, stakeholders must navigate technical, ethical, and regulatory challenges to fully harness its benefits and drive innovation in patient care.

3.4 Challenges and Barriers to Adoption of FL in Medical Case

This section discusses challenges and barriers to the widespread adoption of decentralized approaches, such as the development of standards, education efforts, and addressing ethical concerns.

Federated Learning presents itself as a promising solution to train machine learning models with large and diverse datasets without compromising information confidentiality. This characteristic is significant for healthcare applications, in which Electronic Health Record data is very sensitive by nature and often cannot be easily shared. While there are some successful case studies for the use of FL in production scenarios, various research questions remain open before these architectures become available for widespread usage in healthcare applications [16].

The research in the field of FL applied to EHR data was explored through a review of recent literature that involved several studies [16] proposing and improving methods to ensure the confidentiality of training data. This finding is understandable given the importance of privacy restrictions on individual patients' medical data. Several studies [16] also explore the application of FL on different types of medical data and compare the resulting models with traditional training methods. Other topics addressed include model aggregation methods tailored to medical data needs and architectures to enable FL with multiple medical institutions involved. These insights indicate that FL applied to medical data is an active area of research and that its fundamental aspects are still under development.

In the literature, several studies have aimed to define the challenges associated with the adoption of decentralized approaches, including FL. Significant contributions have been made by authors such as Nicola Rieke et al. [20] and others works like [19]. These studies have identified a number of challenges and barriers that need to be addressed in implementing FL in the medical setting and beyond. The main challenges identified include data heterogeneity, privacy and security, traceability and accountability, and system architecture [19, 20].

- **Data Heterogeneity**: FL may help address certain sources of bias through potentially increased diversity of data sources, but inhomogeneous data distribution poses a challenge for FL algorithms and strategies.
- **Privacy and Security**: Privacy-preserving techniques for FL offer levels of protection that exceed today's current commercially available ML models. However, there is a trade-off in terms of performance, and these techniques may affect the accuracy of the final model.
- **Traceability and Accountability**: Traceability of all system assets including data access history, training configurations, and hyperparameter tuning throughout the training processes is thus mandatory. Addressing concerns related to traceability and accountability in federated learning within the healthcare context requires implementing traceability measures to monitor and record training configurations and model optimizations. This includes tracking changes made to models over time, parameter traceability, and participant contributions, such as computational resources used and data quality for local training.
- **System Architecture**: The administration of such a federation can be realized in different ways. In situations requiring the most stringent data privacy between parties, training may operate via some sort of 'honest broker' system, in which a trusted third party acts as the intermediary and facilitates access to data.
- **Data Integrity and Security**: Ensuring data integrity during communication, designing secure encryption methods to prevent data leakage, and appropriately scheduling node operations to minimize downtime are crucial considerations in FL implementation. In addition, implementing measures for data traceability and accountability is essential to monitor and record training configurations and model.
- **Ethics education and patient involvement**: A proper understanding of the ethical implications of FL is essential for healthcare professionals involved. This requires specific training programs that address issues of data privacy, patient informed consent, and responsibility in the use of FL models. It is critical to educate patients and healthcare professionals about the importance and application of FL in healthcare data management. This educational process should include raising awareness regarding privacy protection and data security, as well as the ethical use of FL technologies in the medical setting. Addressing these ethical concerns is crucial to ensure proper adoption of FL and to promote trust among both users and healthcare professionals.

The adoption of FL in the healthcare sector holds tremendous promise for advancing precision medicine and global healthcare. However, it requires thorough technical considerations and a deep understanding of its impact on all stakeholders involved in the healthcare industry.

FL offers significant benefits in terms of privacy preservation, model accuracy, and data security. By enabling collaborative model training without sharing sensitive data directly, FL empowers healthcare organizations to leverage collective intelligence while maintaining data sovereignty and patient privacy.

Despite its potential, FL implementation faces several challenges, including data heterogeneity, privacy performance trade-offs, trust issues, system architecture complexities, and data integrity and security concerns. Addressing these challenges requires robust infrastructure, standardized protocols, and close collaboration among stakeholders to ensure regulatory compliance and ethical data use [19, 20].

Moving forward, it is crucial for healthcare organizations, clinicians, patients, hospitals, healthcare providers, and medical device manufacturers to work together in navigating the complexities of FL adoption. By doing so, they can unlock the full potential of FL in enhancing healthcare delivery, improving patient outcomes, and driving innovation in the field of medicine [20].

In conclusion, Federated Learning offers a transformative approach to healthcare delivery, ehering a new era of collaborative and data-driven medicine. Through thoughtful consideration of challenges and proactive collaboration, the healthcare industry can harness the power of FL to revolutionize patient care and drive towards a healthier future.

4 Role of the Patient in Decentralized Health Data Management

Decentralized health data management announces a significant paradigm shift in healthcare, placing patients at the center of their own health information ecosystem. This section delves into the important role patients play in this model, emphasizing their involvement in understanding their rights and responsibilities, controlling data access, and actively participating in decision-making processes.

Patients are at the forefront of understanding their rights and responsibilities concerning their health data within decentralized systems. Empowering patients with comprehensive knowledge about privacy regulations, data sharing agreements, and the implications of consenting to data access is essential. By being well-informed, patients can make informed decisions about how their data is managed, shared, and utilized, ensuring their privacy preferences are respected.

Decentralized health data management places control firmly in the hands of patients, allowing them to determine who can access their health information. Through advanced technologies such as blockchain and consent management platforms, patients have the ability to grant or revoke access to their data with precision. This level of control not only fosters trust in the system, but also ensures that patient privacy is upheld, as individuals can manage their data according to their preferences.

Patients are not passive entities within decentralized health data management but active participants in decision-making processes. They contribute to discussions about how their data are used for research, treatment planning, and public health initiatives. By engaging patients in these discussions, healthcare providers and researchers can ensure that their practices are consistent with patient preferences and ethical standards. This collaborative approach empowers patients to play a central role in shaping the ethical and transparent use of health data for better health outcomes.

Patients typically receive treatment locally, but implementing FL on a global scale could ensure high-quality clinical decisions regardless of the treatment location. Specifically, patients in remote areas could benefit from the same high-quality machine learning (ML)-aided diagnoses available in hospitals with a large number of cases. This is particularly beneficial for patients with rare or geographically uncommon diseases, as faster and more accurate diagnoses can lead to milder consequences. Furthermore, FL may lower the barrier for patients to become data donors since they can be reassured that their data remains with their own institution and that data access can be revoked if needed. Thus, FL has the potential to democratize access to advanced medical diagnostics and improve outcomes for patients worldwide, especially those in underserved or remote areas.

In essence, the role of patients in decentralized health data management extends far beyond mere data subjects to become active stewards of their own health information. By being informed about their rights and responsibilities, controlling access to their data, and actively participating in decision-making processes, patients are essential drivers of a patient-centered approach to managing health data. Their participation is fundamental in fostering trust, transparency, and ethical data practices within decentralized health data ecosystems.

5 Future Perspectives for Decentralized Health Data Management

In delineating the future perspectives for decentralized health data management, it is crucial to understand the current context. Presently, health data is often housed in centralized silos controlled by entities such as hospitals, insurance companies, or governmental bodies. This approach poses several significant challenges: patients have little control over their data, data security and privacy are often compromised, access to data is limited, and innovation in the sector is hindered by data centralization. Decentralization emerges as a solution that can transform these problems. By dispersing data in a decentralized network, rather than consolidating them in individual repositories, decentralization has many advantages:

- **Enhanced Control**: Patients gain direct control over their data, empowering them to dictate its usage and sharing, thus reinstating autonomy over their health information.
- **Heightened Privacy and Security**: Distributing data across a decentralized network reduces the risk of breaches and unauthorized access, ensuring patient privacy and fortifying the security of sensitive information.

- **Streamlined Access**: Decentralization simplifies and expedites data access for both patients and healthcare providers, eliminating bureaucratic barriers and fostering seamless and secure information exchange among stakeholders.
- **Fostered Innovation**: The distributed nature of data under decentralization catalyzes innovation within the healthcare sector by granting researchers and developers access to a broader spectrum of data for creating novel healthcare applications and services, consequently enhancing healthcare quality.

Several emerging technologies play an important role in the implementation of decentralized health data management.

- **Federated Learning**: This approach enables machine learning models to be trained on data dispersed across edge devices while protecting the privacy of personal data.
- **Blockchain and Self-Sovereign Identity**: Blockchain technologies and self-sovereign identities establish a secure and decentralized framework for managing personal data, enabling patients to retain control over their data and selectively authorize access.
- **Solid Pods**: Developed within the SOLID project, Solid Pods offer a secure and decentralized means of storing and managing personal data, including health data, empowering patients to maintain data control and selectively share it with authorized entities.

The future prospects of using medical data in a decentralized manner promise to revolutionize the healthcare sector, offering a range of significant advantages while also addressing crucial challenges that must be overcome. Decentralizing medical data could enable quicker and smoother access to healthcare information for both patients and healthcare providers. By eliminating centralized silos, an ecosystem could be created where data is more efficiently and securely accessible. With decentralized data management, patients could have greater control over their privacy. Distributing data across a decentralized network could reduce the risk of privacy breaches and unauthorized access, allowing patients to determine who can access their healthcare information. Decentralization could unlock significant potential for innovation in the healthcare sector. By making data more accessible and usable, it could foster the development of new technologies, applications, and healthcare services that enhance the quality of care and streamline healthcare processes. Decentralization could lead to a reduction in the costs associated with data management and security. By eliminating the need for expensive and complex centralized infrastructures, the financial burden on the healthcare system could be reduced, improving operational efficiency. However, there are also significant challenges that must be addressed: Decentralizing data raises concerns about the security and integrity of the data. It is essential to implement robust measures to protect data from unauthorized access, fraud, and manipulation. Interoperability: Integrating data from different sources and platforms can be complex and require data standardization and interoperability. Without adequate interoperability, decentralization could lead to fragmentation and inefficiencies in data usage. Decentralized data management requires robust governance to ensure compliance with privacy and data security regulations. Developing protocols and regulatory frameworks to regulate the use and exchange of medical data in a decentralized context is necessary. The quality and reliability of data can be compromised in a decentralized environment, where data may be generated and shared from different sources. It's important to implement mechanisms to ensure the integrity and accuracy of data used for clinical and management decisions. The adoption of new decentralized approaches may face resistance from parties accustomed to centralized models. Educating and engaging all stakeholders to promote a smooth transition to decentralized medical data management is crucial. Moving forward, it is imperative for stakeholders in the healthcare industry to collaborate closely, leveraging emerging technologies and addressing challenges proactively to realize the full potential of decentralized health data management in revolutionizing patient care and driving innovation in healthcare delivery.

6 Considerations

The integration of AI into medical cyber-physical systems has revolutionized the management of healthcare data, presenting both opportunities and challenges. Although centralizing healthcare data raises concerns regarding privacy and security, decentralized approaches provide a promising solution by allowing information sharing while protecting individual privacy. This chapter has explored the Solid platform as an example of decentralized healthcare data management, empowering patients with control over their data while enhancing security and interoperability.

Additionally, FL has emerged as a valuable technique for extracting knowledge from distributed data while preserving user privacy, offering applications in early diagnosis, treatment personalization, and patient monitoring. Despite the benefits, challenges persist in adopting decentralized approaches, including the need for standards, education of patients and healthcare professionals, and ethical considerations.

Patients play an important role in decentralized healthcare data management, underscoring the importance of informed involvement in data control decisions. Looking ahead, the future of decentralized data management holds promise for wider adoption, novel applications, and increased public discourse on ethical and societal implications.

Decentralized data management will prioritize techniques like federated learning, efficient algorithms, and enhanced privacy protection, opening new avenues for research in personalized medicine and genomics. Overall, decentralized approaches offer numerous advantages for healthcare data management and will significantly impact AI in medicine.

However, continued development and careful consideration of ethical and legal aspects are imperative to ensure safe, responsible, and beneficial use of healthcare data in the pursuit of improved patient care and medical outcomes.

References

1. Habbal A, Ali MK, Abuzaraida MA (2024) Artificial intelligence trust, risk and security management (AI TRiSM): frameworks, applications, challenges and future research directions. Expert Syst Appl 240:122442. https://doi.org/10.1016/j.eswa.2023.122442
2. Qiu J, Li L, Sun J, Peng J, Shi P, Zhang R, Dong Y, Lam K, Lo FP-W, Xiao B, Yuan W, Wang N, Xu D, Lo B (2023) Large AI models in health informatics: applications, challenges, and the future. IEEE J Biomed Health Inform 27(12):6074–6087. https://doi.org/10.1109/JBHI.2023.3316750
3. Albahri AS, Duhaim AM, Fadhel MA, Alnoor A, Baqer NS, Alzubaidi L, Albahri OS, Alamoodi AH, Bai J, Salhi A, Santamaría J, Ouyang C, Gupta A, Gu Y, Deveci M (2023) A systematic review of trustworthy and explainable artificial intelligence in healthcare: assessment of quality, bias risk, and data fusion. Inf Fusion 96:156–191. https://doi.org/10.1016/j.inffus.2023.03.008
4. Jiménez-Rubio D, García-Gómez P (2017) Decentralization of health care systems and health outcomes: evidence from a natural experiment. Soc Sci & Med 188:69–81. https://doi.org/10.1016/j.socscimed.2017.06.041
5. Cao L (2022) Decentralized AI: edge intelligence and smart blockchain, metaverse, web3, and desci. IEEE Intell Syst 37(3):6–19. https://doi.org/10.1109/MIS.2022.3181504
6. Berners-Lee ST (2024) Solid project. https://solidproject.org/. [Online; Accessed Mar 2024]
7. Sambra AV, Mansour E, Hawke S, Zereba M, Greco N, Ghanem A, Zagidulin D, Aboulnaga A, Berners-Lee T (2016) Solid: a platform for decentralized social applications based on linked data. https://api.semanticscholar.org/CorpusID:49564404
8. Ghayvat H, Sharma M, Gope P, Sharma PK (2022) SHARIF: solid pod-based secured healthcare information storage and exchange solution in internet of things. IEEE Trans Ind Inform 18(8):5609–5618. https://doi.org/10.1109/TII.2021.3136884
9. Ghayvat H, Zuhair M, Shukla N, Kumar N (2024) Healthcare-CT: solid PoD and blockchain-enabled cyber twin approach for healthcare 5.0 ecosystems. IEEE Internet Things J 11(4):6119–6130. https://doi.org/10.1109/JIOT.2023.3312448
10. Chen H (2019) Ubi-care: a decentralized ubiquitous sensing healthcare system for the elderly living support. In: 2019 IEEE Intl Conf on Dependable, Autonomic and Secure Computing, Intl Conf on Pervasive Intelligence and Computing, Intl Conf on Cloud and Big Data Computing, Intl Conf on Cyber Science and Technology Congress (DASC/PiCom/CBDCom/CyberSciTech), pp 543–547. https://doi.org/10.1109/DASC/PiCom/CBDCom/CyberSciTech.2019.00108
11. Ammar N, Bailey JE, Davis RL, Shaban-Nejad A (2021) Using a personal health library-enabled mHealth recommender system for self-management of diabetes among underserved populations: use case for knowledge graphs and linked data. JMIR Form Res 5(3):e24738. https://doi.org/10.2196/24738
12. Harmon DM, Noseworthy PA, Yao X (2023) The digitization and decentralization of clinical trials. Mayo Clin Proc 98(10):1568–1578. https://doi.org/10.1016/j.mayocp.2022.10.001
13. Li L, Fan Y, Tse M, Lin K-Y (2020) A review of applications in federated learning. Comput & Ind Eng 149:106854. https://doi.org/10.1016/j.cie.2020.106854
14. Chen Zhang Yu, Xie HB, Bin Yu, Li W, Gao Y (2021) A survey on federated learning. Knowl Based Syst 216:106775. https://doi.org/10.1016/j.knosys.2021.106775
15. Yu F, Zhang W, Qin Z, Xu Z, Wang D, Liu C, Tian Z, Chen X (2022) Heterogeneous federated learning. arXiv:2008.06767
16. Antunes RS, André da Costa C, Küberle A, Yari IA, Eskofier B (2022) Federated learning for healthcare: systematic review and architecture proposal. ACM Trans Intell Syst Technol 13(4):54. https://doi.org/10.1145/3501813
17. Linardos A, Kushibar K, Walsh S et al (2022) Federated learning for multi-center imaging diagnostics: a simulation study in cardiovascular disease. Sci Rep 12:3551. https://doi.org/10.1038/s41598-022-07186-4

18. Hanif A, Lu C, Chang K, Singh P, Coyner AS, Brown JM, Ostmo S, Chan RVP, Rubin D, Chiang MF, Kalpathy-Cramer J, Campbell JP (2022) Federated learning for multicenter collaboration in ophthalmology: implications for clinical diagnosis and disease epidemiology. Ophthalmol Retin 6(8):650–656. https://doi.org/10.1016/j.oret.2022.03.005
19. Nguyen DC, Pham Q-V, Pathirana PN, Ding M, Seneviratne A, Lin Z, Dobre O, Hwang W-J (2022) Federated learning for smart healthcare: a survey. ACM Comput Surv 55(3):60. https://doi.org/10.1145/3501296
20. Rieke N, Hancox J, Li W, Milletarì F, Roth HR, Albarqouni S, Bakas S, Galtier MN, Landman BA, Maier-Hein K, Ourselin S, Sheller M, Summers RM, Trask A, Xu D, Baust M, Cardoso MJ (2020) The future of digital health with federated learning. NPJ Digit Med 3(1):119. https://doi.org/10.1038/s41746-020-00323-1
21. Rahman MM, Kundu D, Suha SA, Siddiqi UR, Dey SK (2022) Hospital patients' length of stay prediction: a federated learning approach. J King Saud Univ Comput Inf Sci 34(10):7874–7884. https://doi.org/10.1016/j.jksuci.2022.07.006
22. Sehrawat O, Noseworthy PA, Siontis KC, Haddad TC, Halamka JD, Liu H (2023) Data-driven and technology-enabled trial innovations toward decentralization of clinical trials: opportunities and considerations. Mayo Clin Proc 98(9):1404–1421. https://doi.org/10.1016/j.mayocp.2023.02.003

TinyML and Federated Learning for Resource-Constrained Medical Devices

Pietro Fusco, Gennaro Pio Rimoli, and Massimo Ficco

Abstract As the Internet of Things based wearable medical devices become more widely adopted for supporting early diagnosis and remote health monitoring, the amount of data produced by these devices will grow exponentially. Moreover, the high energy consumption for data transmission, the transmission latency, and the privacy and security constraints related to the managed sensitive data can make centralized cloud processing impracticable. Therefore, the need emerges to process data locally through devices capable of making inferences directly on the measurements collected by the sensors worn by patients. On the other hand, the resource constraints of such wearable devices and the computational complexity of exploited Machine Learning algorithms make this approach challenging. In this direction, Tiny Machine Learning and Federated Learning could represent the most appropriate solution for this challenge.

1 Introduction

The Internet of Things (IoT) harnesses the power of edge computing to effortlessly handle data from countless interconnected smart sensors and embedded devices. These IoT devices are strategically placed at the network's edge, boasting minimal memory and processing requirements. They play a pivotal role in IoT ecosystems, relying on edge platforms to gather and transmit data efficiently. Essentially, at the edge, devices collect sensor data, which is then sent for processing either remotely in the cloud or in nearby locations. Edge computing technology not only crunches numbers and stores original data, but also supports distributed computing infrastructure seamlessly. Moreover, edge computing offers several advantages to end-users, including enhanced privacy, security, and reliability. It reduces latency, increases throughput, and ensures better availability, all while delivering prompt responses to services and applications. An intriguing aspect is the collaborative approach among

P. Fusco (✉) · G. P. Rimoli · M. Ficco
University of Salerno, Fisciano, (SA), Italy
e-mail: pfusco@unisa.it

M. Ficco and G. D'Angelo (eds.), *Artificial Intelligence Techniques for Analysing Sensitive Data in Medical Cyber-Physical Systems*, Engineering Cyber-Physical Systems and Critical Infrastructures 12, https://doi.org/10.1007/978-3-031-70775-9_7

smart sensors, edge devices, and the cloud, enabling data processing at the network edge instead of solely relying on cloud resources. This facilitates superior data management, service delivery, data persistence, and content caching. Furthermore, edge computing presents a game-changing opportunity in various applications, such as human-to-machine interaction and smart healthcare. It not only significantly enhances network services, but also automates processes, thus alleviating the burden on network backhaul.

Recent investigations into IoT-edge computing have garnered significant attention for their facilitation of Machine Learning (ML) techniques across numerous applications. However, the hardware earmarked for deployment at the network edge faces severe resource constraints, including limited memory capacity, power consumption, and compatibility issues. These constraints hamper the provisioning of sophisticated, high-end services. On the other hand, the current state of IoT-edge computing falls short of the envisioned cloud-to-embedded paradigm. Despite future expectations, edge computing presently lacks substantial power-saving capabilities and high transmission capacity. This deficiency primarily stems from the disparities between hardware and software technologies, resulting in heterogeneous systems. Consequently, there's a pressing need for holistic and cohesive infrastructures, particularly for tasks like training, updating, and deploying ML models.

Moreover, the architectures tailored for embedded systems vary depending on hardware and software specifics, posing a challenge in establishing a standardized ML architecture for all IoT-edge networks. Additionally, the majority of existing embedded processors are limited to processing sensor data in a generalized manner and executing software applications.

Presently, the vast volume of data generated by multiple devices is routed to the cloud for processing, primarily due to the computationally demanding nature of existing network setups. Utilizing advanced ML models like deep neural networks (DNNs) and deep learning (DL) requires resources, such as graphics processing units (GPUs) and specialized hardware application-specific integrated circuits (ASICs), which consume significant energy and memory capacity.

As a result, there is considerable interest in optimizing ML algorithms to enhance energy efficiency. These developments have given rise to Tiny Machine Learning (TinyML), a concept that implements ML algorithms on miniature devices like IoT-edge devices. TinyML facilitates signal processing on these devices while embedding intelligence, marking a shift away from cloud-based intelligence. Indeed, TinyML represents a rapidly evolving edge computing concept that bridges ML and embedded systems. Overall, TinyML has the potential to enable ultra-low-power and cost-effective systems that demonstrate efficiency and uphold privacy. Additionally, in scenarios with limited connectivity, TinyML can offer on-premise analytics, making it particularly attractive for IoT-based health applications.

Much like other advancements, Federated Learning (FL) has garnered significant interest in recent years. This groundbreaking technology, introduced by Google in 2016, offers a paradigm shift in training ML models collaboratively. Unlike traditional methods that require sending raw data to a central server, FL keeps data on individual devices, fostering a secure and privacy-centric environment.

While FL has found its footing in consumer products, its true potential lies largely untapped within the realm of Industrial IoT (IIoT). The industrial landscape, with its emphasis on data security and distributed operations, presents a perfect breeding ground for FL to flourish. Sensitive production data, limited bandwidth, and security concerns all become non-issues with FL. Imagine a network of industrial machines across a factory floor. Each machine, equipped with FL capabilities, can train a local model on its own sensor data. These models then share only updates, not the raw data itself, with a central coordinator. This collaborative learning process allows for the creation of a powerful, overarching model without compromising individual device privacy. Predictive maintenance, process optimization, and quality control are just a few examples of how FL can revolutionize IIoT. Therefore, by harnessing the collective intelligence of distributed data, FL unlocks insights that were previously inaccessible. This will paves the way for a wave of innovative applications in several domains, including the healthcare sector.

The future of artificial intelligence applied to IoT systems is undeniably intertwined with the advancements in the integration of the TinyML and FL technologies. By fostering secure and collaborative learning, TinyML and FL pave the way for a new era of intelligent and efficient IoT applications. However, further research and development are needed to fully unlock this potential.

2 Understanding TinyML and Federated Learning

TinyML emerges as a versatile machine learning technology equipped to conduct on-device analysis across various sensing modalities, including vision, audio, and speech. Its remarkable attribute lies in its minimal power consumption, making it an ideal fit for embedded resource-constrained edge devices. Moreover, TinyML's compatibility extends to large-scale applications within the Internet of Things (IoT) infrastructure, positioning it as a valuable asset in modern technological frameworks [1, 2].

Nowadays, cloud-based ML systems encounter several challenges, ranging from high power consumption to issues concerning security, privacy, and latency. Consequently, the deployment of pre-installed models on hardware-software platforms, exemplified by Edge Impulse, is gaining attraction [4]. Raw data, reflective of real-world phenomena, is collected by sensors and subsequently processed at a central processing unit (CPU) or microprocessor unit (MPU). These units play a pivotal role in facilitating ML-driven analytics supported by tailored edge-aware ML networks. It's worth noting that edge ML interacts with remote cloud ML systems for knowledge transfer.

The integration of TinyML into such systems holds the promise of significantly enhancing the intelligence of the physical world compared to existing setups [5]. Such a setup empowers edge devices to make critical decisions independently, without relying on edge AI or cloud AI assistance. Importantly, this integration can lead to improvements across various fronts, including energy efficiency, robust data privacy, and reduced latency.

In summary, TinyML represents an integration of hardware, software, and algorithms. In terms of hardware, IoT devices can potentially work in combination with hardware accelerators, but there may be a trade-off between power consumption and computational power. Regarding software, TinyML applications can run on bare-metal Microcontroller Units (MCUs) with or without an operating system like embedded Linux. Furthermore, TinyML systems require new algorithms that produce highly compact models to minimize memory usage. Ultimately, the goal is to optimize machine learning in TinyML systems by efficiently designing software and utilizing high-quality data. This data is then transferred via binary files generated by models trained on larger machines.

Simultaneously, Federated Learning (FL), alternatively referred to as collaborative learning (CL), introduces a decentralized approach to machine learning model training. Unlike traditional methods that involve transferring data from client devices to centralized servers, FL operates by utilizing the raw data stored on edge devices for local model training, thereby bolstering data privacy. This methodology ensures that sensitive information remains on the user's device, safeguarding it from potential breaches or unauthorized access during transmission. The culmination of this process involves collaboratively constructing the ultimate model by aggregating and incorporating the locally trained updates in a shared manner. Through this collaborative effort, FL enables the development of robust and privacy-preserving machine learning models while harnessing the distributed computing power of edge devices.

For example, FL can be a valid approach, which would allow hospitals to collaborate on training powerful AI models for healthcare, all while keeping patient information completely confidential. In summary, the operation would be as follows:

- Hospitals train the models on their local data, without ever sharing the actual patient records with anyone;
- Instead, they only send tiny updates, like the model's learning progress, to a central server;
- The central server combines these updates to create a better overall model, which is then sent back to each involved hospital.

By keeping patient data on-site, FL addresses major privacy concerns. Hospitals can still work together to develop better AI tools without compromising patient confidentiality. This paves the way for exciting advancements in healthcare that benefit both patients and institutions. Figure 1 shows the FL architecture.

FL relies on back-and-forth communication between individual clients (like hospitals) and a central server. Here's a breakdown of this collaborative process:

1. *Requesting the latest knowledge:*

    - Each client starts by reaching out to the central server. It acts like a student seeking new information;
    - The client tells the server it's ready to participate in the training process.

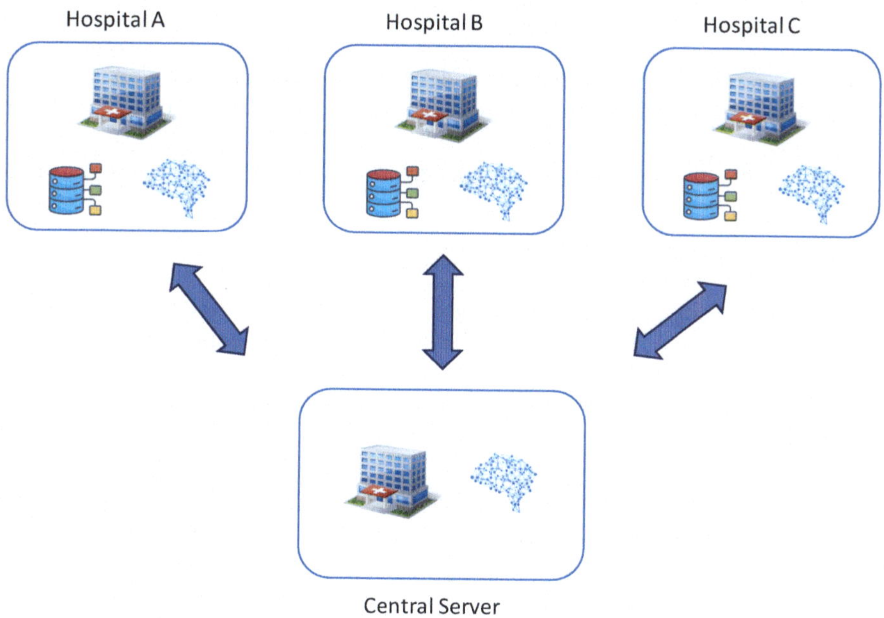

Fig. 1 Federated learning architecture

2. *Obtaining the current model:*
    - In response, the central server sends the client the most up-to-date version of the global model;
    - This model represents the collective learning from all participating clients so far;
    - Think of it as a shared textbook with the latest understanding.
3. *Local training: Applying Knowledge to local data:*
    - Once the client receives the model, it puts it to work;
    - The client trains the model locally using its own private data;
    - This is like the student working on practice problems based on textbook knowledge.
4. *Sharing the learning, not the data:*
    - After the local training is complete, the client doesn't send its raw data back to the server;
    - Instead, it sends an update that captures the improvements the client's data made to the model;
    - Imagine the student sending notes on what concepts were particularly helpful or challenging.

5. *Building a better model together:*

   - The central server receives these updates from all participating clients;
   - It then combines these updates to create a new, improved version of the global model;
   - This is like the teacher gathering insights from all students to refine the class understanding.

6. *The cycle continues:*

   - The central server distributes the improved global model back to all the clients;
   - This initiates a new training round, where each client again refines the model based on its local data;
   - The process continues iteratively, with each round leading to a more robust and generalizable model.

This collaborative dance between clients and the server allows FL to leverage the power of distributed data while keeping sensitive information private.

3 Challenges in Health Management

Health management systems encounter a myriad of challenges, ranging from data privacy concerns to the constraints of computational resources in edge devices, all while grappling with the imperative for real-time decision-making. These challenges underscore the evolving landscape of healthcare, where the integration of cutting-edge technologies intersects with the imperative to safeguard patient privacy and deliver timely, informed care.

Data privacy concerns loom large in the era of digital health records and interconnected medical devices. The sensitive nature of health data necessitates robust safeguards to protect against unauthorized access, breaches, and misuse. Compliance with regulatory frameworks such as HIPAA (Health Insurance Portability and Accountability Act) in the United States adds a layer of complexity, requiring healthcare organizations to implement stringent security measures while ensuring the seamless flow of information for patient care.

Moreover, the proliferation of edge computing and medical devices poses unique challenges in health management systems. These devices, located at the periphery of networks or directly worn by patients. have limited computational and memory capabilities compared to traditional servers or cloud infrastructure. This limitation can hinder the performance of complex algorithms and ML models, necessitating the development of lightweight, efficient solutions tailored to the constraints of tiny computing environments.

In addition, the demand for real-time decision-making in health management further complicates the landscape. From diagnostic assistance to patient monitoring and treatment optimization, timely insights are critical for delivering high-quality care and improving health outcomes. However, achieving real-time responsiveness requires algorithms that can process data rapidly, adapt to dynamic environments, and provide actionable recommendations promptly.

Traditional centralized ML approaches may struggle to address these challenges effectively. Centralized models typically rely on large volumes of centralized data and extensive computational resources, which may not be feasible or practical in edge devices or real-time applications. Furthermore, concerns regarding data privacy and regulatory compliance may limit the availability of centralized data sources, necessitating alternative approaches that prioritize privacy-preserving techniques and decentralized architectures.

In response to these challenges, researchers and practitioners in health management are exploring innovative solutions that leverage decentralized, edge-native ML techniques. FL enables collaborative model training across distributed edge devices without centrally aggregating sensitive data, thereby preserving privacy while harnessing the collective intelligence of diverse datasets. Similarly, tiny computing frameworks such as TensorFlow Lite [7] and PyTorch Mobile [8] offer lightweight, optimized libraries for deploying ML models on resource-constrained devices, enabling real-time inference at the network edge.

In conclusion, addressing the challenges faced by health management systems requires a holistic approach that balances technological innovation with ethical considerations, regulatory compliance, and the practical constraints of real-world implementation. By embracing decentralized, privacy-preserving ML approaches and leveraging the potential of edge computing, healthcare organizations can empower frontline providers with the tools and insights needed to deliver timely, personalized care while safeguarding patient privacy and security.

4 The Role of TinyML FL in Health Management

TinyML Federated Learning is becoming increasingly recognized as a potent solution for tackling the complexities of healthcare data management, particularly on constrained medical devices. This approach leverages lightweight ML models running directly on wearable devices and smart sensors, enabling local data processing. This local processing keeps patient information private and reduces security risks related to the need to transfer and store data on remote data storage. Therefore, TinyML FL offers a win-win scenario for both improved privacy and efficient healthcare data management.

Fueled by COVID-19, the need for continuous cough detection to identify respiratory illnesses has grown significantly. Researchers in [19] have introduced Tiny RespNet, a scalable Convolutional Neural Network (CNN) model that analyzes multiple data types (audio recordings, speech, and demographics) for classification. This model is designed for deployment on Xilinx Artix-7 100t FPGAs, known for their parallel processing capabilities while maintaining low power consumption and high energy efficiency. Tiny RespNet's ability to leverage diverse data allows for effective cough detection and classification of respiratory symptoms, as demonstrated using three datasets.

Researchers explored using deep learning on wearable devices for health diagnosis [10]. They created a TinyML model called TinyDL specifically designed for wearables, focusing on reducing processing time (latency), data transfer (bandwidth), and power consumption. This model, a multi-layered LSTM, analyzed data from the wearable's motion sensors (accelerated data) with accuracy ranging from 75% to 95% for different gestures. However, limitations in the framework prevented processing other data types. Later work [12] addressed these limitations and further explored TinyML's potential for the IoT by reviewing models, compatible devices, data sets, and supporting frameworks. This research highlights the promise of TinyML on-edge devices for overcoming current IoT challenges in healthcare diagnostics.

Beyond health diagnosis, TinyML can also be applied to monitor the well-being of the elderly. Researchers proposed a framework [11] that simplifies adapting models to different platforms, allowing for faster validation and deployment. This framework focuses on body pose estimation, a crucial factor in elderly care. It utilizes algorithms for face landmark and body pose detection, enabling real-time tracking of movement patterns, using a platform like Nvidia Jetson NX, known for its powerful GPUs and deep learning accelerators.

FL offers a groundbreaking approach to healthcare data analysis. It allows healthcare institutions to train machine learning models collaboratively, drawing insights from a vast pool of data spread across different devices. This is achieved without ever needing to share the raw patient information itself, ensuring the utmost privacy. This collaborative nature fosters the development of more powerful and versatile models that can be effectively applied to various health management applications.

5 Applications of TinyML FL in Health Management

TinyML Federated Learning is a revolutionary approach poised to transform health management by harnessing the power of machine learning on resource-constrained devices like wearables and implanted medical equipment. Here are a few contexts where it can be applied:

- *Remote patient monitoring*: TinyML models deployed on wearable devices can continuously monitor vital signs and detect anomalies in real time. FL allows these models to be trained on data from diverse patient populations, improving their accuracy and generalizability.
- *Disease diagnosis*: Edge devices equipped with TinyML can assist healthcare professionals in diagnosing diseases quickly and accurately. FL enables these diagnostic models to learn from decentralized healthcare data, enhancing their performance across different patient demographics and medical conditions.

- *Personalized treatment recommendations*: By analyzing patient data locally on edge devices, TinyML models can generate personalized treatment recommendations tailored to individual health profiles. FL ensures that these models are trained on a wide range of patient data while preserving data privacy.
- *Drug discovery*: TinyML FL can accelerate the drug discovery process by analyzing molecular data from distributed sources. Collaborative model training allows researchers to develop more accurate predictive models for drug efficacy and toxicity without sharing sensitive proprietary data.

6 State of the Art

Artificial intelligence (AI) is a phenomenon that has exploded in recent decades, becoming widely accessible through easy-to-use applications that enable widespread use. AI's potential applications are boundless, all of which depend significantly on extensive data utilization for algorithmic learning and algorithms that demand high hardware utilization for matrix computations. Since 2016, a breakthrough has enabled AI applications to be adapted for devices less computationally powerful, but also consume less power than cloud servers, yet still deliver significant outcomes, known as TinyML algorithms. TinyML remains an emerging concept under investigation in both academic and commercial environments. In the early stage of commercial world, numerous innovative initiatives have been launched, including Google's deployment of its TensorFlow framework, which continues to be prevalent in the sector. Additional instances are PsiKick, now known as Everactive, specializing in predictive maintenance, and Qualcomm's Glance camera module in the vision industry, designed for simple presence detection [13]. On the other hand, the first academic reference began at the TinyML Summit 2019, followed by numerous articles in the next years [3]. The focus was on themes previously explored in the corporate sector, including algorithms that facilitate Computer Vision and Voice Trigger Detection [14]. It also introduced novel computational approaches such as In-Memory Computing (IMC), which achieves 10 to 100 times the energy efficiency and speed of standard processors, and Edge Computing, which relocates processing to the extreme periphery of the network on energy-efficient devices.

ML algorithm can be described as a directed graph. In this graph, nodes represent functions and edges represent data flow and are associated with weights that the algorithm adjusts during training. Each node can receive input from multiple edges, including a special edge called bias, which aids the model in making more accurate predictions by providing an additional parameter. During the inference stage, the trained model is utilized to make predictions or decisions based on new unseen data. The graph's structure remains unchanged, and the weights and biases determined during training are applied to process the new inputs and generate the desired outputs.

TinyML algorithms can be developed using two different strategies [15]. The first approach, which is theoretically remote, involves training the neural network directly on the device. This has several drawbacks, due to the limited memory of the devices on which these algorithms run. To implement this type of algorithm, some studies focus on freezing certain parts of the model [24, 25]. They monitor model updates and try to identify which parts of the network were changed during back-propagation, a step in the training process, but did not significantly improve the overall performance of the model. Consequently, the strategy was to freeze the intermediate weights and update only the biases of each layer or the weights and biases of the final layer to minimize the memory required for training the neural network. Other researchers [30] utilize an incremental Neural Network in which new layers, kept in RAM, are appended to an already trained model stored in ROM, allowing the model to learn new output.

The second method, commonly found in the literature, involves training the model offline or utilizing a pre-trained model [16–21, 23], then deploying a lightweight version on the device, as a C array, through techniques like Pruning, Quantization, Neural Architecture Search, and Hyperdimensional Computing. This approach helps to reduce training durations by leveraging more powerful architectures such as those available in the cloud. However, it requires additional effort to adjust the model post-training and whenever modifications are needed.

In either scenario, the data necessary for training the models needs to be shared externally from the devices. As a result, all data, whether directly or indirectly, is managed by a third party, which may or may not be trusted by the data owners. FL could be used in data sensitive applications (DSA), where safeguarding user privacy is important. The combined use of TinyML and FL is still in the experimental phase [22, 26–28]. Many articles have not concentrated on the use of highly limited IoT development boards like ESP DevKits, STM Nucleo board, Arduino, and Teensy, but instead on more robust Linux-based platforms such as Raspberry Pi. The concept of Federated Learning is the use of multiple clients simultaneously to handle training data while maintaining privacy. Initially, a generic machine learning model is distributed to all client devices, and the collected training data are used independently to train the model weights. The trained weights of each device are then sent to a central entity, often called a server or aggregator. Upon receiving the trained weights from the clients, the server creates a new model by combining the weights using an update metric. This aggregation process ensures that the final model achieves higher accuracy, even when the data distribution among clients is not uniform. Throughout this procedure, optimization methods, especially lossy compression techniques, such as quantization, have been used to enhance the efficiency of data transfer between devices [29]. Alternative research uses blockchain technology to improve security and privacy. In scenarios where the model is attacked or its performance significantly deteriorates, a rollback mechanism is used to revert to a previous version of the model stored in the blockchain, which is uncorrupted [31, 32].

7 Challenges and Future Directions

Although, the excerpt highlights the exciting potential of TinyML FL in revolutionizing health management, it also acknowledges some hurdles that need to be addressed before widespread adoption.

- Challenges:
  - *Communication Bottlenecks*: The current data processing approach involves sending vast amounts of health data from IoT devices (wearables, smart sensors) to the cloud for analysis. TinyML FL proposes processing data on tiny devices themselves before aggregating the insights, reducing data traffic. However, efficient communication protocols are needed to ensure smooth aggregation of these processed insights without delaying decision-making processes.
  - *Data Security*: Health data is highly sensitive. Security measures need to be robust to prevent unauthorized access and breaches.
  - *Regulatory Compliance*: The healthcare industry is heavily regulated. New technologies like TinyML FL need to comply with existing regulations regarding data handling and privacy. Clear guidelines may be needed to navigate this evolving landscape.
- Future Research Directions:
  Despite these challenges, the potential benefits are vast. Here are some promising areas for future research:
  - *Optimizing TinyML Algorithms*: Developing machine learning algorithms that are efficient and accurate for resource-constrained devices at the network edge is crucial. This will enable on-device processing without sacrificing valuable insights.
  - *Secure and Privacy-Preserving FL*: Research on secure communication protocols and encryption techniques is essential to ensure data privacy during the aggregation process in FL. Additionally, anonymization and differential privacy techniques can help protect individual identities while allowing for collective learning.
  - *Decentralized Healthcare Applications*: TinyML FL opens doors for exciting new applications in healthcare. Personalized medicine, where tailored treatment plans are developed based on individual data, and population health management, which focuses on improving the overall health of a community, are just a few examples with immense potential.

By addressing the challenges and exploring these promising research directions, TinyML FL has the potential to revolutionize healthcare delivery, empowering individuals and healthcare providers alike.

8 Conclusions

TinyML Federated Learning is poised to disrupt traditional healthcare management systems by ushering in an era of decentralized and privacy-conscious machine learning on resource-constrained devices at the network edge, including wearables, smart sensors, and even implanted medical devices. This innovative approach unlocks a multitude of benefits, paving the way for a future of:

- *Enhanced Accuracy and Early Detection*: TinyML models, specifically designed for IoT devices, can continuously analyze real-time health data. This continuous monitoring allows for earlier detection of potential health issues and more accurate diagnoses. For instance, analyzing heart rate variability on a smartwatch could provide early warnings of potential heart problems.
- *Personalized Healthcare*: By processing data locally, TinyML FL allows for the development of personalized healthcare solutions. Treatment plans can be tailored to individual needs based on their unique health data, leading to more effective interventions and improved patient outcomes. Imagine a smartwatch that monitors glucose levels and adjusts insulin delivery for diabetic patients.
- *Improved Efficiency and Scalability*: Traditional healthcare systems often rely on centralized data analysis, which can be cumbersome and resource-intensive. TinyML FL distributes the workload across edge devices, reducing the processing burden on central servers. This not only improves efficiency but also allows for easier scaling to accommodate larger patient populations.
- *Reduced Costs*: By enabling on-device processing and localized analysis, TinyML FL can significantly reduce healthcare costs associated with data storage and transmission. Additionally, early detection and personalized interventions can help prevent more expensive complications down the road.
- *Empowering Patients*: TinyML FL empowers patients to take a more active role in managing their health. By having access to real-time insights from their own devices, patients can make informed decisions and collaborate more effectively with healthcare providers.

Beyond the core benefits, TinyML FL also addresses critical challenges in healthcare:

- *Data Privacy Concerns*: Traditional data collection methods raise concerns about data security and privacy. TinyML FL keeps sensitive health data on the user's device, minimizing the risk of breaches and fostering greater trust.

- *Limited Infrastructure*: In resource-constrained regions, access to robust healthcare infrastructure can be a challenge. TinyML FL allows for localized healthcare solutions without relying on centralized servers, improving accessibility to quality healthcare.

In conclusion, TinyML FL holds immense promise for transforming the healthcare landscape. With its focus on decentralization, privacy, and personalized care, it has the potential to empower individuals, revolutionize healthcare delivery models, and ultimately lead to a healthier future for all.

References

1. Ericsson (2019) TinyML as a service and machine learning at the edge. Software available at: https://www.ericsson.com/en/blog/2019/12/tinyml-as-a-service
2. Johnny F, Knutsson Arm F (2021) CMSIS-NN & Optimizations for edge AI. Software available at: https://www.ai.lu.se/2021-03-10
3. TinyML Summit (2019). https://www.tinyml.org/event/summit-2019/
4. Gousev E (2019) Recent progress on tinyml technologies and opportunities. Software available at: https://sites.google.com/g.harvard.edu/tinyml/lectures?authuser=0#h.839rbio9569w
5. Jain P (2019) Edgeml: algorithms for TinyML. Software available at: https://sites.google.com/g.harvard.edu/tinyml/lectures?authuser=0#h.5hc2tcel4ikp
6. Darzidehkalani E, Ghasemirad M, van Ooijen PMA (2022) Federated learning in medical imaging: Part II: methods, challenges, and considerations. J Am Coll Radiol 19(8):975–982. https://doi.org/10.1016/j.jacr.2022.03.016
7. Abadi M et al (2015) TensorFlow: large-scale machine learning on heterogeneous systems. Software available at: www.tensorflow.org
8. Jason A et al (2024) PyTorch 2: Faster machine learning through dynamic python bytecode transformation and graph compilation. In: Proceedings of the 29th ACM international conference on architectural support for programming languages and operating systems, Vol 2 (ASPLOS '24), ACM. https://doi.org/10.1145/3620665.3640366
9. Yang Q, Liu Y, Chen T, Tong Y (2019) Federated machine learning: concept and applications. arXiv:1902.04885
10. Brian C, Md.Shaad M (2021) TinyDL: edge computing and deep learning based real-time hand gesture recognition using wearable sensor. In: 2020 IEEE international conference on e-health networking, application & services (HEALTHCOM), pp 1–6. https://doi.org/10.1109/HEALTHCOM49281.2021.9399005
11. Vuletic M, Mujagic V, Milojevic N, Biswas D (2021) Edge AI framework for healthcare applications. In: Proceedings of the 30th international joint conference on artificial intelligence, Virtual, pp 19–26
12. Alajlan NN, Ibrahim DM (2022) TinyML: enabling of inference deep learning models on ultra-low-power IoT edge devices for AI applications. Micromachines 13(6):851. https://doi.org/10.3390/mi13060851
13. Warden P (2020) TinyML. O'Reilly Media, Sebastopol, CA
14. Sigtia S, Haynes R, Richards HB, Marchi E, Bridle JS (2018) efficient voice trigger detection for low resource hardware. In: Interspeech. https://api.semanticscholar.org/CorpusID:52186997
15. Ray PP (2022) A review on TinyML: State-of-the-art and prospects. J King Saud Univ - Comput Inf Sci 34(4):1595–1623. https://doi.org/10.1016/j.jksuci.2021.11.019
16. Hashir M, Khalid N, Mahmood N, Rehman MA, Asad M, Mehmood MQ, Zubair M, Massoud Y (2023) A TinyML based portable, low-cost microwave head imaging system for brain stroke detection. In: 2023 IEEE international symposium on circuits and systems (ISCAS), pp 1–4. https://doi.org/10.1109/ISCAS46773.2023.10181746

17. Xiaoqun Y, Seonghyeok P, Doil K, Eungjin K, Jaewon K, Woosub K, Yechan A, Shuping X (2023) A practical wearable fall detection system based on tiny convolutional neural networks. Biomed Signal Process Control 86:105325. https://doi.org/10.1016/j.bspc.2023.105325
18. Giordano M, Mayer P, Magno M (2020) A battery-free long-range wireless smart camera for face detection. In: Proceedings of the 8th international workshop on energy harvesting and energy-neutral sensing systems. ACM, New York, NY, pp 29–35. https://doi.org/10.1145/3417308.3430273
19. Tiny RespNet (2020) A scalable multimodal TinyCNN processor for automatic detection of respiratory symptoms. https://api.semanticscholar.org/CorpusID:231400757
20. Flores T, Silva M, Andrade P, Silva J, Silva I, Sisinni E, Ferrari P, Rinaldi S (2022) A TinyML soft-sensor for the internet of intelligent vehicles. In: 2022 ieee international workshop on metrology for automotive (MetroAutomotive), pp 18–23. https://doi.org/10.1109/MetroAutomotive54295.2022.9855110
21. Sudharsan B, Salerno S, Ranjan R (2022) TinyML-CAM: 80 FPS image recognition in 1 kB RAM. In: Proceedings of the 28th annual international conference on mobile computing and networking. ACM, New York, NY, pp 862–864. https://doi.org/10.1145/3495243.3558264
22. Ficco M, Guerriero A, Milite E, Palmieri F, Pietrantuono R, Russo S (2024) Federated learning for IoT devices: enhancing TinyML with on-board training. Inf Fusion 104:102189. https://doi.org/10.1016/j.inffus.2023.102189
23. Xu K, Zhang H, Li Y, Zhang Y, Lai R, Liu Y (2023) An ultra-low power TinyML system for real-time visual processing at edge. IEEE Trans Circuits Syst II: Express Briefs 70(7):2640–2644. https://doi.org/10.1109/TCSII.2023.3239044
24. Cai H, Gan C, Zhu L, Han S (2020) TinyTL: reduce activations, not trainable parameters for efficient on-device learning. arXiv:2007.11622. https://doi.org/10.48550/ARXIV.2007.11622
25. Avi A, Albanese A, Brunelli D (2022) Incremental online learning algorithms comparison for gesture and visual smart sensors. In: 2022 international joint conference on neural networks (IJCNN), pp 1–8. https://doi.org/10.1109/IJCNN55064.2022.9892356
26. Ren H, Anicic D, Runkler TA (2023) TinyReptile: TinyML with federated meta-learning. In: 2023 international joint conference on neural networks (IJCNN), pp 1–9. https://doi.org/10.1109/IJCNN54540.2023.10191845
27. Xiong N, Punnekkat S (2023) Tiny federated learning with bayesian classifiers. In: 2023 IEEE 32nd international symposium on industrial electronics (ISIE), pp 1–6. https://doi.org/10.1109/ISIE51358.2023.10228115
28. Kopparapu K, Lin E, Breslin JG, Sudharsan B (2022) TinyFedTL: federated transfer learning on ubiquitous tiny IoT devices. In: 2022 IEEE international conference on pervasive computing and communications workshops and other affiliated events (PerCom Workshops), pp 79–81. https://doi.org/10.1109/PerComWorkshops53856.2022.9767250
29. Giménez NL, Lee J, Freitag F, Vandierendonck H (2024) The effects of weight quantization on online federated learning for the IoT: A Case Study. IEEE Access 12:5490–5502. https://doi.org/10.1109/ACCESS.2024.3349557
30. Ren H, Anicic D, Runkler TA (2021) TinyOL: TinyML with online-learning on microcontrollers. In: International joint conference on neural networks (IJCNN), pp 1–8
31. Rimoli GP, Boi B, Fusco P, Esposito C, Ficco M (2024) Tiny federated learning with blockchain for privacy and security preservation of MCU-based IoT applications. In: The 6th international conference on blockchain computing and applications (BCCA)
32. Rimoli GP, Fusco P, Ficco M (2024) TinyIDS - An IoT intrusion detection system by tiny machine learning. In: The 24th international conference on computational science and its applications (ICCSA)

The Role of Blockchain in AI-Driven Medical Cyber-Physical Systems

Biagio Boi and Christian Esposito

Abstract The advancement of medical technology and the integration of Cyber-Physical Systems (CPS) have opened up new possibilities for enhanced healthcare services. Electronic Health Record (EHR) aims to standardize the structure of medical data to prevent duplication by providing a high-interoperable framework. However, the growing complexity and sensitivity of medical data in CPS, and their usage in Artificial Intelligence (AI) applications have led to numerous challenges, including data integrity, privacy, and security concerns. Decentralization, particularly blockchain technology, has emerged as a promising solution to address these issues and revolutionize the healthcare industry. In this chapter, we explore applications of blockchain technology in medical CPS, aiming to provide a comprehensive understanding of its potential benefits and limitations within the AI domain. We present a review of existing literature and case studies, highlighting the key advantages of blockchain in the context of medical data management, healthcare supply chains, patient records, and medical Internet of Things (IoT) devices. The incorporation of blockchain in medical IoT devices establishes a robust framework for secure device communication, data integrity verification, and authentication, which are all fundamental properties in AI applications. By leveraging distributed ledger technology, medical IoT networks can mitigate potential cyber threats and protect patient safety. Despite the numerous advantages, this chapter also discusses the challenges and limitations that may hinder the widespread implementation of blockchain for securing AI in the medical domain.

1 Introduction

Smart devices are increasing the efficiency of healthcare by creating new opportunities using continuous sensing and collecting systems able to provide valuable insights. 5G and future 6G networks are paving the way for more interconnected medical systems which are able to offer services based on smart monitoring, remote

B. Boi (✉) · C. Esposito
University of Salerno, Fisciano, SA, Italy
e-mail: bboi@unisa.it

M. Ficco and G. D'Angelo (eds.), *Artificial Intelligence Techniques for Analysing Sensitive Data in Medical Cyber-Physical Systems*, Engineering Cyber-Physical Systems and Critical Infrastructures 12, https://doi.org/10.1007/978-3-031-70775-9_8

control, and remote surgery. A recent study [10] analyzes the innovative Medical 4.0 technologies for healthcare to be applied in order to have an effective CPS involving Artificial Intelligence (AI), Augmented and Virtual Reality (AR/VR), and Telehealth. Despite all these advantages, it is difficult to guarantee privacy and security in these environments, due to the critical and huge amount of data that flows from sensors to servers. Data transferring is one of the most critical points in CPS and needs to be secured as soon as the data are produced. In addition, smart devices have poor capabilities, that make them unable to directly process data or produce well-structured data. Currently, a lot of Public Administration adopts the Electronic Health Record (EHR) (or Electronic Medical Record (EMR)), which is a digital sharable repository that contains all the health data related to a patient. Such a tool has the potential to increase the interoperability of medical data and ensure a more effective medical service. A study on the adoption of EHR [2] confirms advantages such as increased revenue, averted costs, and other benefits that are less tangible, such as improved legal and regulatory compliance, improved ability to conduct research, and so on. Moreover, well-structured data and EHR can be easily used in AI-based decision-making applications [45]. Anyway, considering the importance of these data, the implementation of a fully-compliant EHR system introduces a lot of challenges that need to be addressed. Numerous studies in the literature emphasize issues that hinder the adoption of this technology and affect the development of high-quality AI applications. They can be resumed as follows:

1. Health structures, such as hospitals, usually use not-homogeneous application architectures, and the data extracted are heterogeneous among them [1]. Moreover, it is really difficult to manage real-world data coming from medical context [3], which is usually characterized by poorly documented and poorly structured data such as clinical notes;
2. Patient information is usually stored within a single server, depicting it as a single point of failure. Data centralization includes also a problem related to performances in the access to data which is a critical point in the medical context [4];
3. Privacy concerns are particularly relevant when considering health context; EHR must be capable of ensuring data privacy, and at the same time, not compromising the performance and interoperability of these systems [5].

If digitalization can solve problem related to data structure by applying EHRs, the discussion for privacy and security concerns is more complex. Decentralization is able to solve problems related to centralized servers and privacy, by transferring user data among multiple nodes. Shuaib et al. [11] investigate the adoption of blockchain for solving the problem of EHRs sharing. The advent of blockchain technologies in medical systems is paving the way for a more decentralized way of handling data produced by patients [6]. In such a way data are not stored in a single centralized server but can be stored in multiple nodes which are more resistant to attacks. The flexibility of Blockchain agrees with heterogeneous data that can be stored in a traceable and immutable way.

Despite this new technology sounds like a good thing, able to solve security and privacy issues, there are some aspects that must be carefully evaluated before the adoption. Considering the users' privacy and criticism of produced data, there are multiple aspects to take into consideration for preserving it.

This chapter starts with the analysis of security and privacy requirements that must be taken into consideration when deploying a secure and privacy-aware AI solution. In the third section, an overview of blockchain technology is described in order to understand the main components and different consensus mechanisms. Then, the proposed applications within the medical context are discussed, highlighting the pros and cons of each application. In the fifth section, a discussion on the combination of IoT and blockchain is given, highlighting the limitations. Finally, challenges that come from the usage of blockchain in CPS are discussed.

2 Security and Privacy Requirements

Advantages coming from digitalization are evident in almost all sectors, including the medical one. Despite technologies that can reduce costs and increase the efficiency of the overall processes, it is important to take into consideration security and privacy requirements. The usage of smart sensors usually introduces a huge amount of threats caused by poor security mechanisms implemented on them [7]; but they are not the only point of access, also cloud infrastructure and communication must be secured in order to fulfill the minimal requirements [8]. Figure 1 depicts the typical interactions with data in a cloud-based organization, where patients' data are stored in a healthcare provider and can be consulted by multiple actors. Data access and authorization must be properly implemented and privacy must be taken into consideration. Governments fixed laws for enhancing data protection; two acts must be cited: HIPAA in America and GDPR in Europe. The Health Insurance Portability and Accountability Act (HIPAA) is a federal statute that prohibits sensitive patient health information from being disclosed without the patient's agreement or knowledge. In Europe, General Data Protection Regulation (GDPR) defines the rules that must be taken into consideration for a secure and privacy-preserving system. These rules, considered in a medical context are resumed in Table 1, where some countermeasures are introduced.

Despite these rules being well declared into regulations and a lot of work has been done in proposing frameworks able to adhere to these laws [29]; a lot of companies find it difficult to incorporate and include them in their system. Considering a study conducted in 2020 [9] the EU issued over $500 million in fines, majorly caused by data breaches related to poor secure mechanisms put in place. In addition, when considering privacy and security in the medical context, there are a lot of implications that make patients a direct target for cyber attacks, which are translated into health problems rather than just economic ones. It is possible to introduce some solution to challenges, as well as limitations of existing solutions. R1 is often mitigated through the implementation of authentication protocol. These protocols may rely on lightweight methodologies [27] or leverage unique hardware characteristics [28]. However, both avenues exhibit limitations, with the former compromising

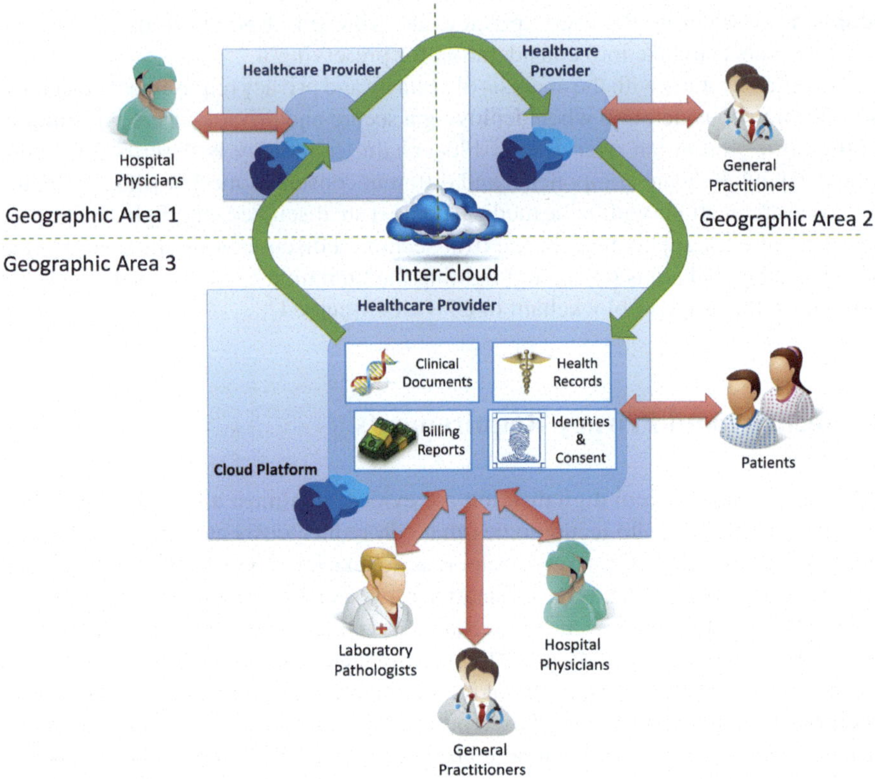

Fig. 1 Typical inter-cloud organization

security and the latter stability. A robust authentication system not only provides authentication but also ensures authorization (referred to as R2). Nevertheless, representing authorization effectively within the medical context proves challenging due to constraints on data formats. In an attempt to address this issue, Takeda et al. [30] propose a mechanism based on the header of HL7 files. Confidentiality (referred to as R3) is often intertwined with authentication, as a sound authentication scheme can inherently uphold confidentiality by facilitating the sharing of encryption keys. For instance, as demonstrated in [29], parties exchange asymmetric keys subsequent to authentication procedures, thereby bolstering confidentiality measures. Auditing techniques (R4) include methods to expose errors and inconsistencies in the medical domain [31]. Typically, an ontology-based approach is used to log actions occurred within a system, but it is hard to secure these logs in an immutable way. The principle of integrity (R5) not only applies to logs system, but also to the data generated by a device or by an entity. Hashing function are used for raw data, while more structured data such as medical images use a watermaking approach [32]. When multiple copies exist for preventing availability (R7), then consistency (R6) must be guaranteed. This requirement is particularly relevant for those systems where multiple devices or entity

Table 1 Privacy and security requirements in medical cyber-physical systems

ID	Name	Description	Typical countermeasure
R1	Authentication	Identification of entity must always be possible in order to prevent identity spoofing	Password, MFA, Secure authentication mechanisms
R2	Authorization	Access to data must be reserved for authorized persons such as patients and doctors	Access Control List (ACL), Role-based Authentication
R3	Confidentiality	Anyone must be able to access users' data without users' consent	Data encryption
R4	Auditing	Monitoring the access to data is a crucial task in the medical context and more in general in CPS	Logging
R5	Integrity	It is referred to the possible modification that can be applied to data produced caused by transmission over corrupted media or attacks from malicious nodes	Anti-tampering
R6	Consistency	Data must be consistent over time. If replication is enabled, all the copies of data must be the same	Database consistency techniques
R7	Availability	Data must be always available, the need for easily accessible data comes from the urgency which is requested in a medical context	Data replication
R8	Patient's consent	Users' consent is needed to control which are the users who agree with tracking their data	Consent management, revokability

interacts with the same data. Finally, patient consent (R8) is a relevant topic when modeling the authorization of users to collect data or to reuse these data for some purpose. Contrary to traditional approaches, decentralization is the main concept that has been considered for the creation of privacy-aware solutions, starting from 2018, the year in which GDPR starts to be effective, an increasing number of works have been published. In what follows we will see how blockchain can be leveraged for enhancing all the described requirements, and which are the main applications currently deployed in the blockchain.

3 Blockchain, What Is It?

A Blockchain is a distributed ledger or database shared among nodes that participate in the network. Before defining the way in which such a system work can be helpful to define which data can be inserted in a distributed ledger. It behaves as a register in which all the operations are stored; for example, it can register the transfer of property of a car or a house. A more generics example can be found in the case of bank wire, where a bank checks for the amount of money stored in the wallet and authorize the operation. Typically these operations are stored in a centralized database owned by Public Administrations or banks thanks to the support of a server. In a decentralized system, the data are stored on nodes that participate in the network.

To be more precise, in a Blockchain, as the word suggests, the data are characterized by transactions that are stored in blocks and then shared among the nodes. As depicted in Fig. 2 a block is composed of multiple transactions. These blocks are progressively concatenated, meaning that newer blocks cannot proceed an already inserted block.

The main advantage of this technology is that older blocks cannot be modified, in fact, a hash value (a value computed on the content of the block, whose modification causes modification of hash value) of each block is computed and a reference to the next block is stored. In such a way whenever the last block is concatenated to a newer block, it cannot be modified anymore. Before adding a new block to the Blockchain, a consensus mechanism is put in place, which is the method able to validate a new block. In the following subsection, an explanation of different consensus mechanisms is discussed.

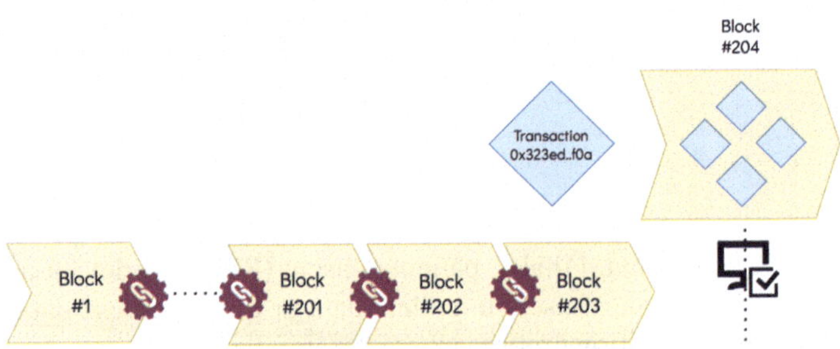

Fig. 2 On the top transaction encapsulation and block creation; on the bottom, the chain of validated blocks

3.1 Blockchain Structure

Depending on the application, different Blockchain exists, on the basis of their visibility and in terms of node participation. It's obvious that if we have a sensible context we do not want to include malicious nodes, which may validate sensible transactions, or if we have a big system we may be interested in scalability.

Classification on visibility:

- *Private Blockchain*—the access to the network is restricted to a specific group of participants; usually the participation is controlled by a central authority or group of entities. Such a Blockchain is designed for organizations where privacy, security, and control are important.
- *Public Blockchain*—the network is open to anyone who wants to view the transactions recorded on the Blockchain. In this case, contrary to the Private Blockchain, there is no authority controlling the network. The adoption of this Blockchain is related to the context where a high level of transparency is needed.

Despite this classification highlights two main categories of Blockchain, it's possible that one is interested in creating a decentralized system where the ledger is publicly available to everyone but the nodes able to validate the transactions are under the control of an authority. It's possible to define the level of access to the consensus: *Permissioned* refers to the Blockchain that controls the nodes able to take part in the consensus; *Permissionless* refers to the Blockchain which does not control the nodes and everyone can take part to the consensus.

3.2 Blockchain Consensus

When a transaction has been sent to the Blockchain network, the nodes must check the validity and authorize it. Since the blockchain leverage decentralization, there is no central authority able to validate or reject a transaction. This verification step takes the name of distributed consensus; the consensus mechanism establishes the way in which the distributed node authorizes a transaction. In a consensus mechanism, nodes that validate a transaction are named *miners*; and usually, they are rewarded for the operation done for validation. Different consensus exists, depending on the operation done for the verification and on the parameter used for making them responsible for the validation; the most popular ones are the following:

- *Proof of Work (PoW)*—the main idea behind *PoW* is an algorithm based on resolving complex mathematical problems such as hash functions for validating a block. A trade-off between complexity and time for the resolution must be established to make new block creation possible in a limited amount of time while guaranteeing strong security against vulnerabilities. *PoW* requires a huge amount of power for solving complex problems, making it poorly efficient.

- *Proof of Stake (PoS)*—in this case, there is no need for complex computation or complex mathematical problems. It aims at reducing the power consumption typical of *PoW* by asking the validators interested in validating an operation an amount of crypto. Such an amount is used for selecting the validators (a higher stake means more probability to be selected as a validator) and is covered by a transaction fee paid from the sender of the validated operation.
- *Proof of Authority (PoA)*—it is based on identity as a stake. Similarly to *PoS*, where the authority can be gained by depositing a higher amount of stake for the validation of transactions; here, the right to validate a transaction can be earned on the basis of a reputation. In this way, validators are incentivized to participate in the transaction process.
- *practical Byzantine Fault Tolerance (pBFT)*—this consensus is majorly used in permissioned blockchain, there the nodes selected for the validation are under the control of an authority. If a node changes behavior over time (namely, it starts to behave maliciously) it is called *byzantine node*. In *pBFT* nodes are ordered with one node being the primary and the others named secondary or backup nodes. The primary node is responsible for the validation, if it fails then the others become primary. In order to prevent malicious validation in *pBFT* is requested that the maximum number of malicious nodes must not be greater than or equal to one-third of all the nodes in the system.

Anyway, with an increasing number of blockchains and distributed ledgers, other consensus mechanisms exist and can be considered, depending on the feature of each blockchain.

4 Blockchain Application in Medical Systems

After explaining blockchain technology and associated aspects, it is critical to grasp its significance within medical CPS. The medical context is one of the major fields where blockchain can be helpful: medical data are one of the most sensitive information and need to be exchanged using secure algorithms and stored in a secure server. Requirements for a privacy-aware and secure medical system have been explained in Sect. 2, here we will clarify which are the major application proposed in the literature leveraging blockchain as means for enhancing users' privacy by acting at different levels of a typical CPS.

4.1 Key Management

Key management is important in all systems, including the medical context. Keys are required to authenticate a user within a system and must be protected from threats. Strong encryption algorithms, such as RSA, are capable of safeguarding data through the utilization of secure asymmetric keys. However, in the event of key compromise, malicious nodes can intercept data either by sniffing the connection or gaining access to a centralized server. Historically, numerous attacks have targeted the Public Key Infrastructure (PKI) and particularly the Certificate Authorities (CAs), entities responsible for issuing key pairs to parties. To mitigate these risks, blockchain technology employs a shared registry that stores all pertinent public-facing information necessary for verification, linking keys to cryptographic wallets [33]. These approaches, similarly to traditional PKI, are composed of two phases [34]. In the first phase, a Trusted Authority (TA) creates a transaction in the blockchain for registering the new device, and associated public key; in the second phase, instead, the authentication processes interact with the blockchain in order to validate the identity of a device (R1). Through the utilization of smart contracts, it becomes feasible to interact with public blockchains to either issue new key pairs or retrieve existing ones from the distributed ledger. Blockchain mechanisms, through the mechanism of distributed consensus, are able to prevent malicious node to attack the entire Decentralized PKI (DPKI). Key Event Receipt Infrastructure (KERI) [40] proposes a rotation strategy to mitigate attacks by frequently changing keys over time. In such an infrastructure, many nodes communicate with one another in order to validate the newly released key. In Sect. 4.3, we will see an actual implementation of this protocol within an identity management system.

4.2 *Data Management*

Data in general, and medical data in particular assumes a fundamental role for the deployment of Machine Learning (ML), Augmented Reality (AR), or Digital Twing (DT) applications [38]. If key management can be executed using a public permissioned blockchain, the same approach cannot be applied to data management. Indeed, medical data cannot be uploaded to a public registry without risking evident data leakage. A typical solution involves leveraging private permissioned blockchains like Hyperledger Fabric [36] to store user data securely. To elaborate, considering that storing data in a public blockchain would compromise confidentiality and privacy, private permissioned blockchains provide controlled access to authorized participants while ensuring data integrity (R5) and confidentiality (R3). A concrete system leveraging this approach has been proposed in [37], where data authenticity and integrity are guaranteed by checking the hash value recorded on the blockchain. The use of blockchain in medical systems can address issues related to data privacy, security, interoperability, and patient engagement. Moreover, smart contracts are able to contain metadata about record ownership, permissions, and data integrity, as evidenced by R2, R3, and R5 [24]. With the evolution of decentralization, data has started to be saved directly on users' devices, utilizing blockchain primarily for integrity checks. These approaches are termed user-centric data management, which solves the problem by excluding relevant data from the blockchain. Dubovitskaya et al. [14] propose a data management system based on EHR that is capable of moving patient data away from a centralized server.

4.3 Identity Management

By storing medical data on a blockchain, we defined a solution to the problem of data leakage through the usage of permissioned blockchain or user-centric data storage, anyway the problem still persist for traditional Identity Management (IdM). Thorugh the use of blockchain it is possible to move identity away from centralized server. In such an approach, patients have full control over their health data and grant access to healthcare providers and researchers as they see fit. This can improve patient privacy and security, as well as facilitate the sharing of data between different healthcare providers and institutions. Self-Sovereign Identity (SSI) is usually used as a synonym of decentralized identity management [21]; it aims to be independent of any Authorization Server (AS) leveraging on Verifiable Credentials (VCs) and cryptographic properties. As shown in Fig. 3, a verifiable data registry contains identifiers and credentials schema, while patients, doctors, and health structures interact with the public registry. When considering secure access to EHRs, SSI plays a crucial role in user authentication. In [23] authors focused on existing solutions, such as Hyperledger Indy and Aries to create a decentralized EHR. Indy is responsible for creating secure credentials leveraging cryptographic property while Aries is able to

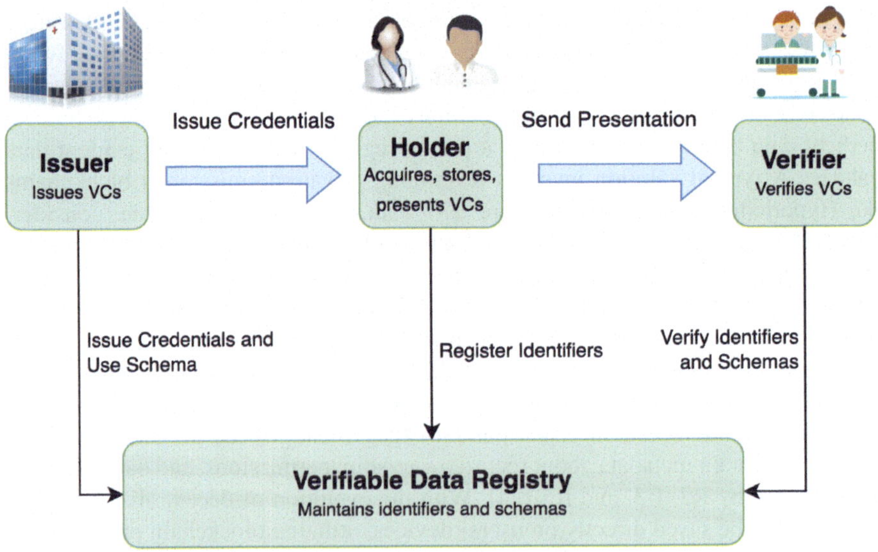

Fig. 3 Medical DIM based on verifiable credentials data model

securely exchange them. The combination of these two hyperledgers paves the way for a more efficient, scalable, and secure IdM.

A different approach has been proposed by HealthID [22], an SSI-based authentication with an Ethereum-based smart contract to facilitate identification and authentication procedures. The HealthID is able to identify each entity (doctors, patients) and leveraging on tokenization, distributes identity among users. Fully decentralized authentication, such as SSI, suffers from the absence of revocation means, as outlined in [41]. KERI can be utilized to safeguard keys employed in SSI-based authentication, showcasing its applicability. Another perspective on revoking identity in a decentralized context based on attestation has been introduced in [42]. Specifically, an alternative decentralized approach to identity management involves the use of Hardware Security Modules (HSM), which holds particular relevance in medical contexts where sensitive parties, such as doctors or physicians, are involved in the system. Further application leveraging attesations will be discussed in Sect. 4.6.

4.4 Supply Chain

Another use case of blockchain in medical systems is to enable secure and transparent tracking of pharmaceutical products and medical devices throughout the supply chain. By using blockchain technology to record the entire supply chain process, from manufacturing to distribution, healthcare providers can verify the authenticity and safety of the products they use, while reducing the risk of counterfeit drugs and devices. Blood transfer is a case of special product, where procedures and storage environments are crucial for endurance over time. BloodMan-Chain [15] tries to automatize the entire management leveraging on a decentralized distributed ledger. It is able to manage all relevant information about blood, making tampering and modification impossible. They also demonstrate data creation and data access using Hyperledger Fabric (permissioned blockchain). Moreover, with the introduction of Non-Fungible Tokens (NFT), traceability and ownership of medical devices are much more secure and transparent [17]. Digital Twin (DT) of the patient is created as a result of transferring the physical characteristics of humans or devices, and related changes in the body to the digital environment [35]. This technology is able to diagnose and predict treatment process suitable for patient with an increased performance, with respect to the traditional methods, which require more complex process for the re-elaboration of data.

4.5 Decentralized Accountability

One of the most important aspects of security aspect is related to accountability, referring to the ability of a system to register every access and action performed over data. In the medical domain, it is hard to account parties for the action performed in the context. As outlined in [43], a huge challenge is related to the organization in that mechanisms for delegation of power from members to representatives may not always be used sufficiently and appropriately, despite that they formally seem to be in place. Considering the special case of medical context, accountability can be extended to resource access taking into consideration the entire CPS. In [18] authors present a real use case based on public healthcare in South Africa, showing how blockchain can help in enhancing transparency and accountability. Timestamp is an irrefutable proof of the time in which the transaction has been done, completely adhering to what is described in R4. Medical data is sensitive and must be kept private even with the introduction of blockchain technology. In [44], the authors propose a novel approach consisting of a public blockchain for managing accountability and Tor hidden services to relay patient data directly to doctors. Both solutions underscore the effectiveness of blockchain for ensuring accountability and emphasize the importance of preserving user privacy by implementing appropriate proxying services.

4.6 Decentralized Attestation

Due to the recent Covid-19 pandemic, attestations received more attention thanks to their natural applicability to immunity certificates and Digital Medical Passport (DMP). Typical certificates depend on a centralized server able to confirm their authenticity by checking signature and revocation lists. As soon as the central servers are violated and compromised, also the certificates are no more trustable. Blockchain guarantees immutability with the concept of block, which makes them not mutable after the validation, awarding R5. As discussed in [19] a good approach is to generate and maintain a certificate on Interplanetary File System (IPFS), while the unique hash obtained by IPFS itself is stored in the blockchain. In this case, any future modification to the certificate is not valid, since it produces a different hash that does not exist on the public ledger.

5 Internet of Medical-Things (IoM-T)

Medical devices have evolved from being isolated and non-networked equipment to devices featuring one-way vendor monitoring, and now to fully networked equipment with bi-directional communications, remote access, wireless connectivity, and software integration. The Internet of Things (IoT) is driving more effective industries, from smart cities in municipalities to advancements in the medical field. The Internet of Medical Things (IoMT) includes Cyber-Physical Systems (CPS) and smart sensors aimed at enhancing lives and reducing medical costs through efficient connectivity, particularly with the advent of 6G networks. Many IoMT devices rely on machine-to-machine communication, minimizing the need for human intervention.

The application of Machine Learning (ML) in the IoT context has spurred its expansion, reducing costs while improving the quality of predictions. IoMT is a pivotal technology for a more efficient healthcare industry, and blockchain offers the privacy and security features necessary for medical systems. However, the limited memory and battery capacity of IoMT devices make it challenging to perform complex tasks typically required by applications. Blockchain can mitigate potential threats in this context [25].

Combining these technologies is challenging due to their differing requirements and constraints. Esposito et al. [13] suggest a promising approach: using off-chain storage for healthcare data, while storing immutable hashes on-chain to verify authenticity. Despite the limitations and constraints of IoMT devices, real-time computation is difficult, and transmitting all generated data is impractical due to security and network load concerns. Edge Computing addresses these issues by performing computations within the network, enabling aggregation and analysis tasks to transmit only relevant data. Aggregation and anonymization by edge devices help ML models comply with regulations. Numerous literature solutions combine edge computing and blockchain. Abdellatif et al. [12] propose MEdge-Chain, leveraging both technologies for efficient medical data exchange. This system enables multiple local networks of hospitals to communicate and securely store user medical data in a blockchain network. Rocha et al. [46] analyze distributed intelligence, presenting architectures split into multiple layers: cloud, edge, and end devices. IoT devices and sensors are revolutionizing data collection, and are promising technologies for the development of real-time applications. In the following section, we will analyze the limitations and challenges of the usage of blockchain within a medical context.

6 Challenges

The higher the number of hospitalized persons, the higher the quantity of data produced by them; motivating the need for a highly scalable solution to produce reliable ML models. Such solutions must be able to guarantee privacy and security in order to avoid possible data leakage. Blockchain is a crucial technology for the enhancement of security and privacy thanks to its immutability and related consensus mechanisms. As introduced, different blockchains based on their visibility exist and need to be accurately chosen in order to prevent possible attacks on the node's validity. One of the major concerns in the application of Blockchain is related to the cost and to the technology to adopt [39]. The integration of blockchain within the Internet of Medical-Things is still at the beginning, and future improvements can be made in order to increase security in these last devices.

GDPR and HIPAA define the law to be taken into consideration when developing a privacy-aware solution, and a lot of frameworks are automatizing the extraction of requirements from textual policies, facilitating this task. Decentralized identity can be applied also within the medical context, by moving away from centralized servers' patient data and enhancing the concept of decentralized authentication, which does not require any authentication server. On the contrary, the Decentralization of users' data is not always possible due to intrinsic characteristics of the blockchain (ex. node locations), which goes in contrast with some GDPR principles; but with proper architecture and design phases, it is a good solution to scalability.

EHR standardizes the representation of medical data, enhancing interoperability among applications. Despite such a standardization, the efficiency of the collection phase must be secured and efficiently sped up. Edge computing is partially solving the problem but more work must be done for making EHR interoperable with low-power devices [20]. Web of Things (WoT) is standardizing the entire IoT context, leveraging on three concepts: properties, actions, and events. Such standardization is necessary for making also IoM-T more interoperable enabling the so-called M2M communication.

7 Conclusion and Future Development

Overall, the use of Blockchain in medical systems has the potential to improve the quality of care, reduce costs, and give patients a more reliable system of handling health data. However, the adoption of Blockchain technology in the healthcare industry is still in its early stages, it requires further research and development to fully realize its potential.

Power and memory constraints are huge challenges in CPS, which pose some limitations on the expansion of the blockchain in this context, also considering privacy and security requirements. Anyway, a solution is represented by energy harvesting techniques that are able to increase the battery capacity by harvesting energy from natural sources like solar, thermal, or RF. A special energy source in a medical context is characterized by ultrasound-based energy harvesters for implantable devices [26] that are able to collect enough power to secure and transmit sensitive data without the need for a battery.

References

1. Nicolai B, Tallarico S, Pellegrini L, Gastaldi L, Vella G, Lazzini S (2022) Blockchain for electronic medical record: assessing stakeholders' readiness for successful blockchain adoption in health-care. Meas Bus Excel 27(1):157–171. https://doi.org/10.1108/MBE-12-2021-0155
2. Menachemi N, Collum TH (2011) Benefits and drawbacks of electronic health record systems. Risk Manag Healthc Policy. Taylor & Francis, pp 47–55
3. Knevel R, Liao KP (2023) From real-world electronic health record data to real-world results using artificial intelligence. Ann Rheum Dis 82(3):306–311. BMJ Publishing Group Ltd
4. Rao KR, Naganjaneyulu S (2021) Permissioned healthcare blockchain system for securing the EHRs with privacy preservation. Ingénierie des Systèmes d'Information 26(4)
5. Tertulino R, Antunes N, Morais H (2023) Privacy in electronic health records: a systematic mapping study. J Public Health. Springer, pp 1–20

6. Merlo V, Pio G, Giusto F, Bilancia M (2022) On the exploitation of the blockchain technology in the healthcare sector: a systematic review. Expert Syst Appl. Elsevier, pp 118897
7. Huang M, Liu A, Wang T, Huang C (2018) Green data gathering under delay differentiated services constraint for internet of things. Wirel Commun Mob Comput 2018. Hindawi
8. Sun W, Cai Z, Li Y, Liu F, Fang S, Wang G (2018) Security and privacy in the medical internet of things: a review. Secur Commun Netw 2018:1–9. Hindawi Limited
9. Daigle B, Khan M (2020) The EU general data protection regulation: an analysis of enforcement trends by EU data protection authorities. J Int'l Com & Econ 1. HeinOnline
10. Haleem A, Javaid M, Singh RP, Suman R (2022) Medical 4.0 technologies for healthcare: Features, capabilities, and applications. Internet Things Cyber Phys Syst 2:12–30. Elsevier
11. Shuaib K, Abdella J, Sallabi F, Serhani MA (2022) Secure decentralized electronic health records sharing system based on blockchains. J King Saud Univ Comput Inf Sci 34(8):5045–5058. Elsevier
12. Abdellatif AA, Samara L, Mohamed A, Erbad A, Chiasserini CF, Guizani M, O'Connor MD, Laughton J (2021) Medge-chain: Leveraging edge computing and blockchain for efficient medical data exchange. IEEE Internet Things J 8(21):15762–15775. IEEE
13. Esposito C, De Santis A, Tortora G, Chang H, Choo K-KR (2018) Blockchain: a panacea for healthcare cloud-based data security and privacy? IEEE Cloud Comput 5(1):31–37. IEEE
14. Dubovitskaya A, Baig F, Xu Z, Shukla R, Zambani PS, Swaminathan A, Jahangir MM, Chowdhry K, Lachhani R, Idnani N, and Others (2020) ACTION-EHR: Patient-centric blockchain-based electronic health record data management for cancer care. J Med Internet Res 22(8):e13598. JMIR Publications Toronto, Canada
15. Le TH, Trong PN, Gia KH, Vo HK, Huong LH, Dang KT, Van HL, Huu NH, Huyen TN, Nguyen TA, and Others (2022) BloodMan-chain: a management of blood and its products transportation based on blockchain approach. In: International conference on parallel and distributed computing: applications and technologies, pp 199–209. Springer
16. Essefi I, Boussi Rahmouni H, Solomonides T, Fethi Ladeb M (2022) HIPAA controlled patient information exchange and traceability in clinical processes. In: 2022 IEEE 9th international conference on sciences of electronics, technologies of information and telecommunications (SETIT), pp 452–460. IEEE. https://doi.org/10.1109/SETIT54465.2022.9875865
17. Gebreab SA, Hasan HR, Salah K (2022) NFT-based traceability and ownership management of medical devices. IEEE Access 10:126394–126411. IEEE
18. Ndayizigamiye P, Dube S (2019). Potential adoption of blockchain technology to enhance transparency and accountability in the public healthcare system in South Africa. In: 2019 international multidisciplinary information technology and engineering conference (IMITEC), pp 1–5. IEEE
19. Namasudra S, Sharma P, Crespo RG, Shanmuganathan V (2023) Blockchain-based medical certificate generation and verification for IoT-based healthcare systems. IEEE Consum Electron Mag 12(2):83–93. https://doi.org/10.1109/MCE.2021.3140048
20. Sauer CM, Chen, L-C, Hyland SL, Girbes A, Elbers P, Celi LA (2022). Leveraging electronic health records for data science: common pitfalls and how to avoid them. Lancet Digit Health 4(12):e893–e898. Elsevier
21. Satybaldy A, Hasselgren A, Nowostawski M (2022) Decentralized identity management for E-Health applications: state-of-the-art and guidance for future work. Blockchain Healthc Today 5. Partners in Digital Health, LLC
22. Javed IT, Alharbi F, Bellaj B, Margaria T, Crespi N, Qureshi KN (2021) Health-ID: a blockchain-based decentralized identity management for remote healthcare. Healthcare 9:712. MDPI
23. Manoj T, Makkithaya K, Narendra V (2022) A blockchain based decentralized identifiers for entity authentication in electronic health records. Cogent Eng 9(1):2035134
24. Mayer AH, da Costa CA, da Rosa RR (2020) Electronic health records in a Blockchain: a systematic review. Health Inform J 26(2):1273–1288
25. Ratta P, Kaur A, Sharma S, Shabaz M, Dhiman G (2021) Application of blockchain and internet of things in healthcare and medical sector: applications, challenges, and future perspectives. J Food Qual 2021:1–20

26. Ballo A, Bottaro M, Grasso AD (2021) A review of power management integrated circuits for ultrasound-based energy harvesting in implantable medical devices. Appl Sci 11(6):2487
27. El-Meniawy N, Rizk MRM, Ahmed MA, Saleh M (2022) An authentication protocol for the medical internet of things. Symmetry 14(7):1483
28. Shao X, Guo Y, Guo Y (2022) A PUF-based anonymous authentication protocol for wireless medical sensor networks. Wirel Netw 28(8):3753–3770
29. Xu Z, Xu C, Liang W, Xu J, Chen H (2019) A lightweight mutual authentication and key agreement scheme for medical internet of things. IEEE Access 7:53922–53931
30. Takeda T, Ueda K, Nakagawa A, Manabe S, Okada K, Mihara N, Matsumura Y (2017) A document-based EHR system that controls the disclosure of clinical documents using an access control list file based on the HL7 CDA header. In: MEDINFO 2017: precision healthcare through informatics, pp 1238–1238
31. Zheng L, He Z, Wei D, Keloth V, Fan J-W, Lindemann L, Zhu X, Cimino JJ, Perl Y (2020) A review of auditing techniques for the unified medical language system. J Am Med Inform Assoc 27(10):1625–1638
32. Su G-D, Chang C-C, Lin C-C (2020) Effective self-recovery and tampering localization fragile watermarking for medical images. IEEE Access 8:160840–160857
33. Pal O, Alam B, Thakur V, Singh S (2021) Key management for blockchain technology. ICT Express 7(1):76–80
34. Garg N, Wazid M, Das AK, Singh DP, Rodrigues JJPC, Park Y (2020) BAKMP-IoMT: design of blockchain enabled authenticated key management protocol for internet of medical things deployment. IEEE Access 8:95956–95977
35. Erol T, Mendi AF, Doğan D (2020) The digital twin revolution in healthcare. In: 2020 4th international symposium on multidisciplinary studies and innovative technologies (ISMSIT), pp 1–7
36. Rouhani S, Butterworth L, Simmons AD, Humphery DG, Deters R (2018)MediChainTM: a secure decentralized medical data asset management system. In: IEEE international conference on internet of things (iThings), vol 2018, pp 1533–1538
37. Yang Q, Liu Q, Lv H (2020) A decentralized system for medical data management via blockchain. J Internet Technol 21(5):1335–1345
38. Zahid A, Poulsen JK, Sharma R, Wingreen SC (2021) A systematic review of emerging information technologies for sustainable data-centric health-care. Int J Med Inform 149:104420
39. Sternberg HS, Hofmann E, Roeck D (2021) The struggle is real: insights from a supply chain blockchain case. J Bus Logist 42(1):71–87
40. Smith SM (2019) Key event receipt infrastructure (keri). arXiv:1907.02143
41. Boi B, De Santis M, Esposito C (2024) Enhancing security in user-centered authentication using KERI. In: The 32nd Euromicro international conference on parallel, distributed, and network-based processing (PDP 2024)
42. Boi B, Esposito C, Seo JT (2024) Ethereum Attestation Service as a solution for the revocation of hardware-based password-less mechanisms. In: The 39th ACM/SIGAPP symposium on applied computing, pp 553–559
43. Kuo T-T, Pham A (2022) Detecting model misconducts in decentralized healthcare federated learning. Int J Med Inform 158:104658
44. Ali MS, Vecchio M, Putra GD, Kanhere SS, Antonelli F (2020) A decentralized peer-to-peer remote health monitoring system. Sensors 20(6):1656
45. Hossain E, Rana R, Higgins N, Soar J, Barua PD, Pisani AR, Turner K (2023) Natural language processing in electronic health records in relation to healthcare decision-making: a systematic review. Comput Biol Med 155:106649
46. Rocha A, Monteiro M, Mattos C, Dias M, Soares J, Magalhães R, Macedo J (2024) Edge AI for internet of medical things: a literature review. Comput Electr Eng 116:109202

Artificial Intelligence Applications in Healthcare Security

Salvatore D'Antonio and Federica Uccello

Abstract This chapter explores the capabilities of Artificial Intelligence (AI) within cybersecurity in healthcare. In particular, the integration of Artificial Intelligence AI into traditional security monitoring framework is discussed, introducing AI-based Security Monitoring approaches, combining rule-based SIEM with Machine Learning (ML) techniques, such as Association Rule Mining (ARM) for improved detection and Explainable AI (xAI) for transparency. Adaptive Security Monitoring, addressing dynamic cyber threats, is discussed through Concept Drift Detectors. Finally, the application of AI-based Security Monitoring in healthcare is highlighted. Various use cases are presented, such as the protection of patient data, medical devices, and network. The integration of AI in access control and threat intelligence is examined to ensure confidentiality, prevent unauthorized access, and maintain healthcare system integrity.

1 Introduction

The rapid technological advancement of the last decade has brought numerous advantages. Increased connectivity, cloud computing, and the Internet of Things (IoT) have enhanced efficiency, collaboration, and accessibility. However, with these advancements, novel threats have arisen, bringing significant challenges to cybersecurity and, consequently, to the safety of individuals and the reliability of critical systems. Cybersecurity threats do not only concern data integrity but also hold disruptive potential, especially considering the interconnected nature of virtually any infrastructure. A violation towards a single target might end up compromising numerous others, in a disastrous cascading effect.

Present Address:
F. Uccello (✉)
Linköping University, Linköping, Sweden
e-mail: federica.uccello@liu.se; federica.uccello@assegnista.uniparthenope.it

S. D'Antonio · F. Uccello
University of Parthenope, Naples, Italy

M. Ficco and G. D'Angelo (eds.), *Artificial Intelligence Techniques for Analysing Sensitive Data in Medical Cyber-Physical Systems*, Engineering Cyber-Physical Systems and Critical Infrastructures 12, https://doi.org/10.1007/978-3-031-70775-9_9

In this complex landscape, recognizing the central role of data is essential. Data is among the most important assets of the digital age. It enables businesses to make informed decisions, governments to deliver essential services, and individuals to connect and collaborate across vast distances. Moreover, the value of data extends beyond its immediate utility, serving as a catalyst for innovation and economic growth. On one hand, the collection and analysis of vast amounts of personal data raise concerns about privacy. Data breaches, unauthorized surveillance, and the misuse of personal information can compromise individuals' privacy and lead to identity theft or other forms of cybercrime [1]. Additionally, the increasing reliance on interconnected systems and digital infrastructure has given rise to a proliferation of cyber threats. Cyberattacks including malware, phishing, and ransomware, pose significant risks to individuals, organizations, and critical systems [2].

Within the healthcare domain in particular, the risk connected to malicious exploitation can extend far beyond privacy concerns. In such a scenario, people safety and even humans lives are on the line, especially with the shift towards digitally-enabled healthcare services. With the proliferation of electronic medical records and inherent security flaws in Information Technology (IT) processes, hospitals became easy target for cybercriminals. Ransomware is among the most concerning threats in this area [3]: among the most notable ones, The WannaCry epidemic from 2017 was one of the largest ransomware attacks in history, affecting hospital facilities in over 150 countries worldwide [4].

With the outbreak of the COVID-19 pandemic, new threats emerged as well. The pandemic posed immense strain on healthcare systems worldwide, while also exposing vulnerabilities to cyber-attacks within the healthcare sector. As medical providers struggled with the unprecedented demand for clinical services, cybercriminals took the opportunity to launch sophisticated attacks on public and private organizations, including healthcare institutions and academic centers. INTERPOL's warning in April 2020 underscored the urgent need for robust cybersecurity measures to protect sensitive healthcare data and ensure uninterrupted delivery of essential services during these challenging times [5].

The Health Insurance Portability and Act (HIPAA) journal also mandated the protection and confidential handling of sensitive patient information within the healthcare sector. The latest update, as of January 22, 2024, reveals a concerning trend of escalating data breaches in the healthcare industry. Despite efforts to mitigate breaches through improved digital record-keeping and encryption, hacking incidents and ransomware attacks have surged, comprising nearly 80% of reported breaches in 2023. Moreover, the severity of breaches has intensified, with 133 million records exposed in 2023 alone, marking a milestone in data security vulnerabilities within the healthcare sector [6].

Thus, it is clear that safeguarding the integrity, confidentiality, and availability of data is vital, as any compromise can have far-reaching consequences for individuals, organizations, and society as a whole. Therefore, robust security measures, including effective Security Monitoring, are essential to protect data from unauthorized access, manipulation, and exploitation.

2 Security Monitoring

Security Monitoring focuses on identifying and mitigating security threats such as intrusions, data breaches, and malicious activities.

This section explores the foundations of Security Monitoring, considering two key elements in particular, which are crucial for implementing effective security strategies: Security Information and Event Management Systems (SIEM) and Intrusion Detection Systems (IDS).

2.1 Security Information and Event Management Systems

Within real-time security monitoring, the role of SIEM systems has been subject of discussion and research in the last few years [7–9], highlighting the importance of early detection in today's complex landscape.

According to the United States National Institute of Standards and Technology (NIST), a SIEM is defined as an "application that provides the ability to gather security data from information system components and present that data as actionable information via a single interface" [10].

SIEM systems stand as core elements in the realm of cybersecurity, as they serve as the nerve center of security operations by collecting and analyzing log data from various sources. Their capabilities extend beyond data aggregation, enabling detection of patterns, anomalies, and potential security incidents.

The primary functionalities of SIEM include centralized log management, as they consolidate logs from various sources across the network or system. This unified view enables efficient log analysis, aiding in the detection of security events and incidents.

Moreover, the real-time monitoring capabilities of SIEM systems empower organizations to promptly respond to emerging threats. By continuously analyzing incoming data, these systems can identify and address security incidents in their early stages.

Incident Response is a further core capability of SIEM systems, which provide comprehensive insights into the nature and scope of security incidents, facilitating a more informed and timely response to mitigate potential risks.

An high level architecture of SIEM is schematized in Fig. 1, and it is described in the following [11]. The principal modules include:

- Data Collection: SIEM systems must be able to collect data from various sources such as network devices, servers, and applications. For this scope, a dedicated module equipped with various probes is fundamental. Through the probes, diverse data streams are channeled into the SIEM ecosystem, ensuring comprehensive coverage of the monitored environment.

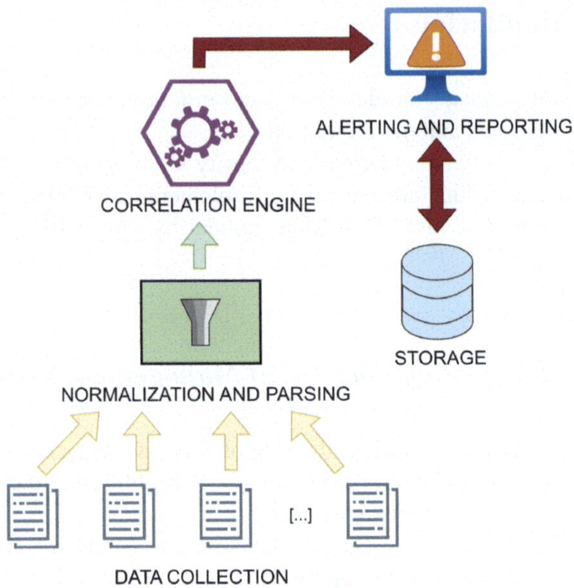

Fig. 1 High level architecture of traditional rule-based SIEM. The main building blocks and the data flow are shown

- Normalization and Parsing: Due to the heterogeneous formats and structures of the information collected by the various sources, it becomes imperative for it to undergo normalization and parsing processes. Through these procedures, data is harmonized into a standardized format, enabling consistency and compatibility for further analysis. This ensures that all the data ingested by the SIEM follows a standard format.
- Correlation Engine: This component is the kernel of the architecture, as it is in charge of correlating data from different sources to identify patterns and potential security incidents. According to the implemented correlation logic, in fact, the correlation engine is able to detect anomalies an suspicious behaviors.
- Alerting and Reporting: SIEM systems generate alerts and reports based on the detected anomalies. These alerts serve as early warning signals, notifying security personnel of potential security breaches or suspicious activities warranting further investigation. Additionally, SIEM are often equipped with dashboards displaying the most relevant metrics, to further support incident management, and providing valuable insights regarding the situational picture of the incident.
- Storage: In addition to real-time analysis and alerting, SIEM systems are equipped with databases for historical data. This is a key component to support and facilitate compliance, forensic analysis, and the refinement of threat detection capabilities over time.

2.2 Intrusion Detection Systems

While SIEM systems provide a broader context for Security Monitoring, IDS operate on the front lines, acting as specialized probes that delve into specific activities to detect potential threats with precision.

IDS are able of monitoring network or system activities for any signs of malicious behaviors, following distinct approaches [12, 13]. The approaches can be classified in three broad categories [14]: Signature-based IDS, Anomaly-based IDS, and Specification-based IDS, also known as Stateful protocol analysis.

- Signature-based IDS: These systems rely on predefined patterns or signatures of known threats. A well-know example include common anti-virus programs. This category of IDS proves high effectiveness in identifying well-established attack vectors, however it may struggle with unknown or sophisticated attacks.
- Anomaly-based IDS: These systems establish a baseline of normal behavior by monitoring the system or application throughout a training phase. Then, the IDS raise alerts when deviations are detected. This category is effective in identifying previously unknown threats, however it requires accurate tuning to avoid the generation of high false positives.
- Specification-based IDS: These systems are capable of tracking and correlating protocol states, such as pairing requests with corresponding replies. While these IDS resemble anomaly-based ones, the latter rely on preloaded network or host-specific profiles, whereas Specification-based ones rely on vendor-developed generic profiles tailored to specific protocols.

Table 1 shows a comparison between the three categories, highlighting pros and cons of each.

While SIEM systems provide a holistic view of the overall security landscape, IDS systems offer a more focused and specialized approach. The two technologies

Table 1 Advantages and limitations of the most common IDS approaches

	Signature-based	Anomaly-based	Specification-based
Advantages	Most simple and effective method to detect known attacks. It enables detailed contextual analysis	Effective to detect zero day vulnerabilities. Useful in detecting privilege abuse attacks	Aware of the protocol states. It can detect unexpected sequences of commands
Limitations	Less effective against novel threats. Less aware of states and protocols. Signature and patterns need to be frequently checked and updated	Can suffer from low training accuracy. Can not be used during the training. High false positive rate	Resource consuming due to protocol state tracing and examination. It struggles with detecting attacks looking like genuine protocol behaviors

complement each other by combining broad contextual awareness with targeted, in-depth analysis. SIEM can benefit from the data collected by IDS as a part of their security probe set. In the following, AI-enhanced Security Monitoring is discussed, along with the introduction of a novel security solution.

3 AI-Based Security Monitoring

As the landscape of cybersecurity threats continues to evolve, there is a growing imperative to enhance traditional Security Monitoring technologies to meet the challenges caused by sophisticated and dynamic attacks [15, 16].

The integration of Artificial Intelligence (AI) techniques into traditional SIEM and IDS has shown potential to enhance the current state of such systems and bring multiple advantages in terms of pattern recognition, real-time detection, and continuous learning [17].

Traditional SIEM and IDS systems, while robust, may face limitations in adapting to the constantly changing nature of cyber threats. AI technologies, particularly Machine Learning (ML), can significantly enhance the capabilities of classic monitoring systems in various ways. For instance, a next-gen SIEM framework leveraging ML is proposed in [18].

AI is exploited in [19] to create a robust SIEM system, meant to protect Industrial Internet of Things (IoT) applications.

Other research investigates the application of Deep Reinforcement Learning (DRL) in optimizing post-alert incident responses in SIEM systems [20].

The ML-IDS proposed in [21] aims at detecting IoT network attacks, with a focus on privacy challenges. In the same application domain, [22] proposes an IDS enhanced with a hybrid Convolutional Neural Network (CNN) model.

The study conducted in [23] proposes a packet-based ML model with deferred decision and a single packet integration classifier for network intrusion detection.

The solution proposed in [24] consists of a two-stage hybrid methodology for intrusion detection using ML.

A classification-based NIDS using big data analytics tools and deep learning approaches is proposed in [25].

Through a Hybrid Security Monitoring methodology, it is possible to address the aforementioned limitations of traditional approaches. Rule-Based SIEM can be augmented using ML algorithm and techniques in order to improve the accuracy of the detection, providing transparency and trust, and enabling adaptive Security Monitoring.

The main functionalities of such enhanced approach are depicted in Fig. 2 and described in the following subsections.

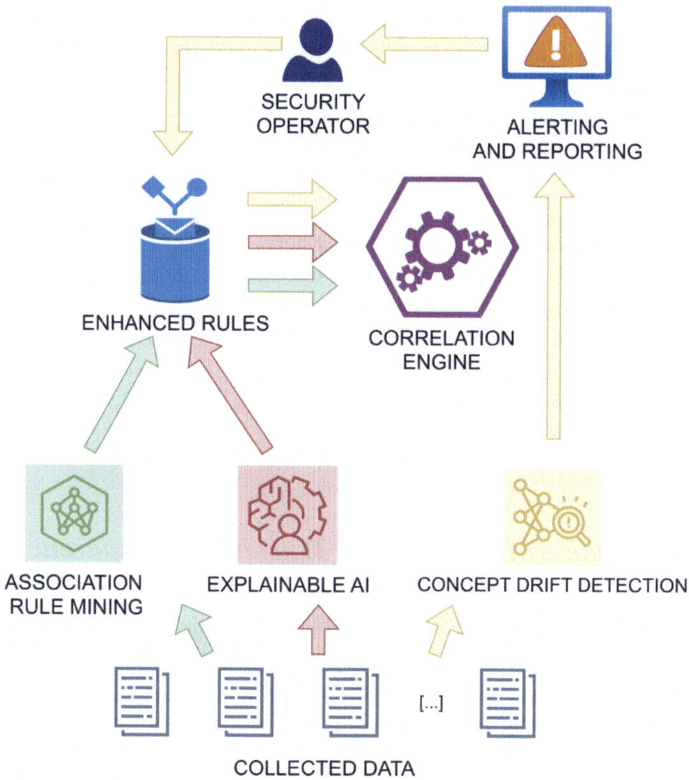

Fig. 2 High level architecture of AI-enhanced Security Monitoring. The main building blocks and the data flow are shown

3.1 Association Rule Mining for Improved Detection

The majority of traditional SIEM systems are equipped with a Rule-based Correlation Engine. Such component allows the security personnel to define static correlation and detection rules to identify malicious pattern by combining data from multiple sources, including IDS, and raising alert when anomalies are observed.

ML algorithms, such as Association Rule Mining (ARM), hold potential to refine the correlation and detection criteria used by SIEM systems.

In [26], the authors investigate the integration of associative rules into SIEM systems to improve the detection of cyber incidents. The proposed approach incorporates fuzzy sets theory and data mining techniques to construct associative rules based on the analysis of cyber incidents.

In [27], sequential rule mining is employed in the analysis of intrusion detection alerts. The main goal is to reduce the volume of alerts received by security professionals.

The work presented in [28] addresses the common issue of false positive alerts generated by intrusion detection systems. The study proposes a novel network security IDS framework that leverages Modified Frequent Pattern (MFP-Tree) with the K-means algorithm in detecting several kinds of attacks.

The authors of [29] introduce a method for mining association rules from multi-source logs to identify various intrusion behaviours in cloud computing platforms.

Among the others, APRIORI [30] is a classic algorithm for association rule discovery, that efficiently identifies correlations between attributes in the dataset. The APRIORI algorithm is applied to preprocessed datasets, resulting in the identification of association rules that describe relationships between different features. These rules provide insights into the behaviors and patterns associated with network traffic. As ARM identifies hidden patterns and relationships in data, it enables the definition of highly accurate correlation criteria which might be impossible to implement with simple observation and domain knowledge.

3.2 Explainable AI for Accuracy, Trust and Transparency

Explainable Artificial Intelligence (xAI) [31] can provide insights in enhancing the accuracy, trustworthiness, and transparency of ML models and classifiers.

In the context of Security Monitoring, xAI can support the understanding of decision-making processes, providing more trust and transparency.

Transparency is a key aspect of security systems. In the realm of cybersecurity, where the consequences of false positives or false negatives can have a major impact, xAI enables the creation of transparent models by providing interpretable explanations for decisions. Security analysts and stakeholders can analyze these explanations, fostering confidence in the accuracy of the system's predictions.

While multiple xAI methods have been proposed in the literature, the most common include feature importance, rule-based explanations, and counterfactual explanation. These methods can help cybersecurity professionals understand how ML models are making decisions, as well as identifying potential biases in the models and supporting debugging problems that might arise. Additionally, xAI can be employed to identify the most important features for the model's predictions.

In the realm of IDS, [32] introduces an advanced IDS utilizing ML ensemble methods like Decision Trees (DT), Random Forest (RF), and Support Vector Machines (SVM), achieving promising results. It incorporates xAI with the LIME algorithm to enhance model transparency.

In [33], the critical role of trust management in IDS is explored, emphasizing the need for transparency in ML models utilized in the cybersecurity domain. It discusses the challenges of interpreting "black-box" AI models and the importance of xAI in enhancing trust by enabling human experts to comprehend the model's decision-making process.

The paper [34] introduces an IDS employing ML techniques like DT, RF, and xAI on real-world SDN data. The evaluation covers various intrusion scenarios, achieving high accuracy.

In [35], an approach for detecting DDoS attacks using xAI is proposed. The method focuses on identifying attack behaviours in network traffic flows without analyzing packet payloads.

The work carried out in [36] aims at contributing to rigorous xAI-equipped IDS for DDoS attacks.

Similarly to ARM, xAI can be used to support security professionals in the definition of correlation and detection rules with a high degree of accuracy. Enhanced rules can be mined through several well-know xAI algorithms, including SHapley Additive exPlanations (SHAP) and ANCHOR. The interpretability of xAI allows for the identification of critical features that contribute to specific outcomes.

3.3 Enabling Adaptive Security Monitoring

Within data analysis, a significant issue is represented by the temporal alteration in the statistical characteristics of a target variable, also known as Concept Drift. In the context of cybersecurity, this can translate to changes in the patterns of malicious activities.

Traditional systems may struggle to cope with the dynamic nature of data in real-time Security Monitoring, as the slight pattern change might stay unnoticed long enough for attack to be executed and cause irreversible damage.

The potential of Concept Drift Detection in Intrusion Detection has also been explored in past research. The approach shown in [37] employs an incremental feature extraction method, and a stable sub-classifier generation module to achieve high efficiency and adaptability to various real-world drifts, with promising experimental results.

Rajeswari et al. [38] propose a methodology for large-scale intrusion detection. The approach employs data cleaning and normalization, with a dedicated algorithm detecting drift and a component classifying data as normal or attack. Comparative analysis with various techniques is also provided.

A methodology for monitoring changes in network traffic from individual source nodes based on concept drift is proposed in [39]. The framework applies ML techniques to dynamically define lower and upper bounds for each node, effectively discriminating abrupt changes from gradual changes in traffic magnitude. Real-world data testing validates the methodology's effectiveness.

Beside the aforementioned applications, AI can enable adaptive Security Monitoring through the incorporation of Concept Drift Detectors, which can be implemented in a supervised or unsupervised fashion.

In a supervised approach, the model is trained on labeled data, and it learns to recognize changes in the target variable with the help of predefined labels.

On the other hand, unsupervised approaches do not rely on labeled data for detecting concept drift. These methods analyze the data distribution over time and identify shifts or changes without the need for predefined labels.

In real-life cybersecurity applications, where obtaining labeled data is impractical, unsupervised approaches prove to be more suitable for real-time detection of Concept Drift. Among the unsupervised Concept Drift Detectors, Page-Hinkley Test (PHT) [40] and Adaptive Windowing (ADWIN) [41] can be employed within a hybrid Security Monitoring.

In such context, Concept Drift Detection allows the system to self-adapt once the defined correlation criteria display becomes obsolete, or to raise a warning prompting the security operator to revise the rules in sight for unexpected performance decay in the detection.

4 AI for Healthcare Protection

In the complex landscape of Healthcare, AI can offer innovative solutions to address concerns related to data and system security [42, 43]. The intersection of AI and healthcare security is particularly relevant in addressing the balance between providing personalized and efficient patient care, and ensuring confidentiality and integrity of sensitive medical information.

The inherent complexity of modern healthcare, characterized by vast datasets, interconnected medical devices, and intricate network infrastructures, necessitates advanced technological solutions. AI, with its capacity to analyze massive datasets, predict outcomes, and respond in real-time, has emerged as a supportive tool in such a scenario.

Ensuring the security and confidentiality of patient data is mandatory in healthcare practices. AI's integration into access control mechanisms adds layers of sophistication, automating authentication processes and monitoring access patterns to detect anomalies in real-time. This not only prevents unauthorized access but also mitigates potential breaches, aligning healthcare practices with privacy regulations and increasing trust in the security of medical information.

AI-powered threat intelligence also becomes instrumental in continuously monitoring, analyzing, and responding to potential cyber threats. From early detection to proactive mitigation, AI can ensure that healthcare organizations are equipped to defend against cyber attacks, safeguard patient data, and maintain the integrity of critical healthcare systems.

This section explores the evolving role of AI in the cybersecurity domain for medical applications. In the following, possible application areas for AI-enhanced Security Monitoring discussed in Sect. 3 are presented, including protection of patient data, medical devices, and network.

4.1 Patient Data Security

In healthcare, AI can be a powerful tool in safeguarding patient data security.

To begin with, AI-powered security solutions offer scalability and adaptability, which are crucial in the dynamic landscape of modern healthcare environments. As the volume and complexity of patient data continue to grow, AI enables healthcare organizations to efficiently manage and protect this information. In addition, it can help ensuring compliance with regulatory requirements such as the Health Insurance Portability and Accountability Act (HIPAA) in the United States or the General Data Protection Regulation (GDPR) in the European Union.

AI-enhanced IDS and SIEM constitute a main component in such a scenario, by continuously monitoring access patterns to patient records. Going beyond traditional rule-based approaches, AI-driven SIEM and IDS dynamically learn normal patterns of user access, enabling them to discern subtle deviations indicative of potential security threats. By establishing a baseline of normal user behavior, the system can rapidly identify anomalies, flagging instances that fall outside the established norms.

This not only allows for the rapid detection of unauthorized access but also positions healthcare providers to proactively face potential breaches before they escalate. In fact, the efficient and quick detection of anomalous behaviour can trigger immediate responses.

By analyzing vast amounts of data in real-time, AI can uncover patterns and correlations that might elude human analysts, thereby enhancing the overall resilience of healthcare cybersecurity infrastructure.

4.2 Medical Device Security

Another critical aspect of modern healthcare infrastructure is the protection of medical devices.

In fact, such devices are becoming increasingly interconnected as a part of the complex IoT-Healthcare, and AI is able to offer scalability and adaptability to meet their evolving complexity.

AI techniques can enable robust defense against potential cyber threats, minimizing the risk of disruptions to patient care. It provides capabilities to enhance medical device security by enabling predictive analytics and proactive threat mitigation. By analyzing historical data and identifying emerging patterns, AI algorithms can predict potential security risks beforehand.

AI-enhanced IDS and SIEM can actively monitor the behavior of medical devices in real-time, identifying anomalies that may indicate unauthorized access or manipulation. In this way, it is not only possible to protect patient safety by ensuring the proper functioning of medical devices but also to safeguards the integrity of the broader healthcare infrastructure.

One of the primary concerns in medical device security is preventing unauthorized access. AI algorithms can learn and adapt to the normal usage patterns of medical devices, making it possible to distinguish between legitimate and potentially malicious activities. This adaptive capability is crucial in thwarting unauthorized access attempts, whether they originate from internal sources or external cyber threats.

4.3 Network Security

Network security is a further area of concern to take into consideration when protecting healthcare systems. In fact, network security plays a crucial role in ensuring the confidentiality, integrity, and availability of critical healthcare data and services.

The integration of various technologies in healthcare, ranging from Electronic Health Records (EHR) systems to medical imaging devices, has significantly expanded the attack surface for potential security breaches. A compromise in network security can have far-reaching consequences, impacting the functionality of medical devices, compromising patient data, and potentially disrupting the overall delivery of healthcare services. As such, addressing network security has become a major concern in healthcare facilities.

The scalability and adaptability of AI-driven security solutions make them well-suited for the dynamic and complex nature of healthcare networks. AI-driven Security Monitoring offers advanced capabilities in monitoring and analyzing network traffic patterns. One of the major advantages of AI-enhanced SIEM and IDS is their ability to not only detect potential threats but also to automate and facilitate efficient mitigation actions. When a security incident is identified, AI systems can trigger automated responses, isolating affected areas of the network, blocking suspicious traffic, or notifying security personnel for further investigation.

This rapid response mechanism is essential in containing and neutralizing cyber threats, enabling the definition of mitigation actions and preventing cascade effects.

References

1. Khan FS, Kim JH, Moore RL, Mathiassen L (2019) Data breach risks and resolutions: a literature synthesis
2. Palleti VR, Adepu S, Mishra VK, Mathur A (2021) Cascading effects of cyber-attacks on interconnected critical infrastructure. Cybersecurity 4:1–19
3. Thamer N, Alubady R (2021) A survey of ransomware attacks for healthcare systems: risks, challenges, solutions and opportunity of research. In: 2021 1st Babylon international conference on information technology and science (BICITS), pp 210–216
4. Chen Q, Bridges RA (2017) automated behavioral analysis of malware: a case study of WannaCry ransomware. In: 2017 16th IEEE international conference on machine learning and applications (ICMLA), pp 454–460
5. INTERPOL (2020) Preventing crime and protecting police: INTERPOL's COVID-19 global threat assessment. https://www.interpol.int/en/News-and-Events/News/2020/Preventing-

crime-and-protecting-police-INTERPOL-s-COVID-19-global-threat-assessment. Accessed on 04 June 2024
6. HIPAA Journal on healthcare data breach statistics (2024). https://www.hipaajournal.com/healthcare-data-breach-statistics/. Accessed on 04 June 2024
7. Podzins O, Romanovs A (2019) Why SIEM is irreplaceable in a secure IT environment?. In: 2019 open conference of electrical, electronic and information sciences (eStream), pp 1–5
8. González-Granadillo G, González-Zarzosa S, Diaz R (2021) Security information and event management (SIEM): analysis, trends, and usage in critical infrastructures. Sensors 21(14):4759
9. Rosenberg M, Schneider B, Scherb C, Asprion PM (2023) An adaptable approach for successful siem adoption in companies. arXiv:2308.01065
10. Johnson A, Dempsey K, Ross R, Gupta S, Bailey D et al (2011) Guide for security-focused configuration management of information systems. NIST Spe Publ 800(128):16–16
11. Sheeraz M, Paracha MA, Haque MU, Durad MH, Mohsin SM, Band SS, Mosavi A (2023) Effective security monitoring using efficient SIEM architecture. Hum-Centric Comput Inf Sci 13:1–18
12. Raiyn J (2014) A survey of cyber attack detection strategies. Int J Sec Appl 8(1):247–256
13. Khraisat A, Gondal I, Vamplew P, Kamruzzaman J (2019) Survey of intrusion detection systems: techniques, datasets and challenges. Cybersecurity 2(1):1–22
14. Liao H-J, Lin C-HR, Lin Y-C, Tung K-Y (2013) Intrusion detection system: a comprehensive review. J Netw Comput Appl 36(1):16–24
15. Cucu C, Cazacu M (2019) Current technologies and trends in cybersecurity and the impact of artificial intelligence. Int Sci Conf eLearn Softw Educ 2:208–214
16. Campfield M (2020) The problem with (most) network detection and response. Netw Sec 2020(9):6–9
17. Zhang Z, Ning H, Shi F, Farha F, Xu Y, Xu J, Zhang F, Choo KK (2022) Artificial intelligence in cyber security: research advances, challenges, and opportunities. Artif Intell Rev 1–25
18. Ban T, Takahashi T, Ndichu S, Inoue D (2023) Breaking alert fatigue: AI-assisted SIEM framework for effective incident response. Appl Sci 13(11):6610
19. Radoglou-Grammatikis P (2023) SecureCyber: an SDN-enabled SIEM for enhanced cybersecurity in the industrial internet of things. IEEE COMSOC MMTC Commun - Front 18(2):2023
20. Alturkistani H, El-Affendi MA (2022) Optimizing cybersecurity incident response decisions using deep reinforcement learning. Int J Electr Comput Eng 12(6):6768
21. Saheed YK, Abiodun AI, Misra S, Holone MK, Colomo-Palacios R (2022) A machine learning-based intrusion detection for detecting internet of things network attacks. Alexandria Eng J 61(12):9395–9409
22. Smys S, Basar A, Wang H et al (2020) Hybrid intrusion detection system for internet of things (IoT). J ISMAC 2(04):190–199
23. Kim T, Pak W (2022) Real-time network intrusion detection using deferred decision and hybrid classifier. Futur Gener Comput Syst 132:51–66
24. Rao KN, Rao KV, PVGD PR (2021) A hybrid intrusion detection system based on sparse autoencoder and deep neural network. Comput Commun 180:77–88
25. Al S, Dener M (2021) STL-HDL: a new hybrid network intrusion detection system for imbalanced dataset on big data environment. Comput Sec 110:102435
26. Subach I, Mykytiuk A, Methodology of formation of fuzzy associative rules with weighted attributes from SIEM database for detection of cyber incidents in special information and communication systems. Inf Technol Sec 11(1) (20)
27. Husák M, Bajtoš T, Kašpar J, Bou-Harb E, Čeleda P (2020) Predictive cyber situational awareness and personalized blacklisting: a sequential rule mining approach. ACM Trans Manag Inf Syst (TMIS) 11(4):1–16
28. Sivanantham S, Mohanraj V, Suresh Y, Senthilkumar J (2023) Association rule mining frequent-pattern-based intrusion detection in network. Comput Syst Sci Engi 44(2):1403–1414

29. Bounou M, Alouani A, El Azzaoui A (2022) Combining artificial intelligence techniques to improve anomaly detection in SIEM systems. In: Proceedings of the 2022 4th international conference on intelligent communication technologies and virtual mobile networks (ICICV), pp 120–126
30. Hegland M (2007) The apriori algorithm–a tutorial. Mathematics and computation in imaging science and information processing, pp 209–262
31. Gunning D, Stefik M, Choi J, Miller T, Stumpf S, Yang GZ (2019) XAI-explainable artificial intelligence. Sci Robot 4(37):eaay7120
32. Patil S, Varadarajan V, Mazhar SM, Sahibzada A, Ahmed N, Sinha O, Kumar S, Shaw K, Kotecha K (2022) Explainable artificial intelligence for intrusion detection system. Electronics 11(19):3079
33. Mahbooba B, Timilsina M, Sahal R, Serrano M (2021) Explainable artificial intelligence (XAI) to enhance trust management in intrusion detection systems using decision tree model. Complexity 1–11:2021
34. Karna SK, Paudel P, Saud R, Bhandari M Explainable prediction of features contributing to intrusion detection using ML algorithms and LIME
35. Kalutharage CS, Liu X, Chrysoulas C, Pitropakis N, Papadopoulos P (2023) Explainable AI-based DDOS attack identification method for IoT networks. Computers 12(2):32
36. Zhou Q, Li R, Xu L, Nallanathan A, Yang J, Fu A (2022) Towards explainable meta-learning for DDoS detection. arXiv:2204.02255
37. Wang X (2022) ENIDrift: a fast and adaptive ensemble system for network intrusion detection under real-world drift. In: Proceedings of the 38th annual computer security applications conference, pp 785–798
38. Rajeswari PV, Shashi M, Rao TK, Rajya Lakshmi M, Kiran LV (2022) Effective intrusion detection system using concept drifting data stream and support vector machine. Concurr Comput: Pract Exp 34(21):e7118
39. Rajeswari PVN, Shashi M (2019) Concept-drift based identification of suspicious activity at specific IP addresses using machine learning. Int J Recent Technol Eng 8(3):6651–6655
40. Page ES (1954) Continuous inspection schemes. Biometrika 41(1/2):100–115
41. Bifet A, Gavalda R (2007) Learning from time-changing data with adaptive windowing. In: Proceedings of the 2007 SIAM international conference on data mining, pp 443–448
42. Telo J (2017) AI for enhanced healthcare security: an investigation of anomaly detection, predictive analytics, access control, threat intelligence, and incident response. J Adv Anal Healthcare Manag 1(1):21–37
43. Adil M, Khan MK, Farouk A, Jan MA, Anwar A, Jin Z (2022) Ai-driven EEC for healthcare IoT: Security challenges and future research directions. IEEE Consum Electron Mag

The manufacturer's authorised representative in the EU is Springer Nature Customer Service Centre GmbH, Europaplatz 3, 69115 Heidelberg, Germany. If you have any concerns regarding our products, please contact ProductSafety@springernature.com

Printed and bound by CPI Group (UK) Ltd, Croydon, CR0 4YY
25/09/2026
02222892-0018